HOW DID THE

First Stars and Galaxies Form?

HOW DID THE

First Stars and Galaxies Form?

ABRAHAM LOEB

PRINCETON UNIVERSITY PRESS

PRINCETON AND OXFORD

Published by Princeton University Press, 41 William Street,
Princeton, New Jersey 08540
In the United Kingdom: Princeton University Press,
6 Oxford Street, Woodstock, Oxfordshire OX20 1TW

ISBN: 978-0-691-14515-0
ISBN (pbk.): 978-0-691-14516-7

Library of Congress Control Number: 2010923723

British Library Cataloging-in-Publication Data is available

This book has been composed in Garamond
Printed on acid-free paper ∞
press.princeton.edu

Typeset by S R Nova Pvt Ltd, Bangalore, India
Printed in the United States of America

10 9 8 7 6 5 4 3 2 1

To my parents, Sara and David,
who gave me life,
and my three women, Ofrit, Klil, and Lotem,
who made it worthwhile

CONTENTS

PREFACE

This book captures the latest exciting developments concerning one of the unsolved mysteries about our origins: *how did the first stars and galaxies light up in an expanding Universe that was on its way to becoming dark and lifeless?* I summarize the fundamental principles and scientific ideas that are being used to address this question from the perspective of my own work over the past two decades.

Most research on this question has been theoretical so far. But the next few years will bring about a new generation of large telescopes with unprecedented sensitivity that promise to supply a flood of data about the infant Universe during its first billion years after the Big Bang. Among the new observatories are the James Webb Space Telescope (JWST)—the successor to the Hubble Space Telescope—and three extremely large telescopes on the ground (ranging from 24 to 42 meters in diameter), as well as several new arrays of dipole antennae operating at low radio frequencies. The fresh data on the first galaxies and the diffuse gas in between them will test existing theoretical ideas about the formation and radiative effects of the first galaxies, and might even reveal new physics that has not yet been anticipated. This emerging interface between theory and observation will constitute an ideal opportunity for students considering a research career in astrophysics or cosmology. With this in mind, the book is intended to provide a self-contained introduction to research on the first galaxies at a level appropriate for an undergraduate science major or a scientist with nonspecialist background. Many of the nontechnical chapters are also suitable for the educated general public.

Various elements of the book are based on a cosmology class I have taught over the past decade in the Astronomy and Physics Departments at Harvard University. Other parts relate to overviews I wrote over the past decade in the form of five review articles (three with Rennan Barkana) and five popular-level articles (one with Avery Broderick and one with T. J. Cox). Where necessary, selected references are given to advanced papers and other review articles in the scientific literature.

The writing of this book was made possible thanks to the help I received from a large number of people. First and foremost, I am grateful to my parents, Sara and David, who supported my journey through life with unconditional love and understanding. I also thank the many graduate students and senior collaborators with whom I had fun learning about the contents of this book, including Dan Babich, John Bahcall, Rennan Barkana, Laura Blecha, Avery Broderick, Volker Bromm, Renyue Cen, Benedetta Ciardi, Mark Dijkstra, Daniel Eisenstein, Claude-André Faucher-Giguère, Richard Ellis, Steve Furlanetto, Zoltan Haiman, Lars Hernquist, Loren Hoffman, Bence Kocsis, Shri Kulkarni, Piero Madau, Joey

Muñoz, Ramesh Narayan, Ryan O'Leary, Jerry Ostriker, Jim Peebles, Rosalba Perna, Jonathan Pritchard, Fred Rasio, Martin Rees, George Rybicki, Dan Stark, Max Tegmark, Anne Thoul, Hy Trac, Ed Turner, Eli Visbal, Eli Waxman, Stuart Wyithe and Matias Zaldarriaga. Special thanks go to Claude-André Faucher-Giguère, Joey Muñoz, Tony Pan, Jonathan Pritchard, and Greg White for their careful reading of the book and detailed comments, to Joey Muñoz and Hy Trac for their help with several figures, and to Donna Adams for her assistance with the LaTex file and the illustrations. Finally, I would like to particularly thank the love of my life, Ofrit Liviatan, who established the foundations on which I stood while writing this book, and our two daughters, Klil and Lotem, who inspired my thoughts about the future.

A. L.
Lexington, MA

HOW DID THE

First Stars and Galaxies Form?

1

PROLOGUE: THE BIG PICTURE

1.1 In the Beginning

As the Universe expands, galaxies get separated from one another, and the average density of matter over a large volume of space is reduced. If we imagine playing the cosmic movie in reverse and tracing this evolution backward in time, we can infer that there must have been an instant when the density of matter was infinite. This moment in time is the "Big Bang," before which we cannot reliably extrapolate our history. But even before we get all the way back to the Big Bang, there must have been a time when stars like our Sun and galaxies like our Milky Way* did not exist, because the Universe was denser than they are. If so, *how and when did the first stars and galaxies form?*

Primitive versions of this question were considered by humans for thousands of years, long before it was realized

---

*A **star** is a dense, hot ball of gas held together by gravity and powered by nuclear fusion reactions. A **galaxy** consists of a luminous core made of stars or cold gas surrounded by an extended halo of *dark matter* (see section 2.7).

that the Universe expands. Religious and philosophical texts attempted to provide a sketch of the big picture from which people could derive the answer. In retrospect, these attempts appear heroic in view of the scarcity of scientific data about the Universe prior to the twentieth century. To appreciate the progress made over the past century, consider, for example, the biblical story of Genesis. The opening chapter of the Bible asserts the following sequence of events: first, the Universe was created, then light was separated from darkness, water was separated from the sky, continents were separated from water, vegetation appeared spontaneously, stars formed, life emerged, and finally humans appeared on the scene.* Instead, the modern scientific order of events begins with the Big Bang, followed by an early period in which light (radiation) dominated and then a longer period dominated by matter, leading to the appearance of stars, planets, life on Earth, and eventually humans. Interestingly, the starting and end points of both versions are the same.

1.2 Observing the Story of Genesis

Cosmology is by now a mature empirical science. We are privileged to live in a time when the story of genesis (how the Universe started and developed) can be critically explored by direct observations. Because of the finite time it takes light to travel to us from distant sources, we can

---

*Of course, it is possible to interpret the biblical text in many possible ways. Here I focus on a plain reading of the original Hebrew text.

WMAP 5-year

−200   T(μk)   +200

Figure 1.1. Image of the Universe when it first became transparent, 400 thousand years after the Big Bang, taken over five years by the Wilkinson Microwave Anisotropy Probe (WMAP) satellite (http://map.gsfc.nasa.gov/). Slight density inhomogeneities at the level of one part in $\sim 10^5$ in the otherwise uniform early Universe imprinted hot and cold spots in the temperature map of the cosmic microwave background on the sky. The fluctuations are shown in units of μK, with the unperturbed temperature being 2.73 K. The same primordial inhomogeneities seeded the large-scale structure in the present-day Universe. The existence of background anisotropies was predicted in a number of theoretical papers three decades before the technology for taking this image became available.

see images of the Universe when it was younger by looking deep into space through powerful telescopes.

Existing data sets include an image of the Universe when it was 400 thousand years old (in the form of the cosmic microwave background in figure 1.1), as well as images of individual galaxies when the Universe was older than a billion years. But there is a serious challenge: in between these two epochs was a period when the Universe was dark, stars had not yet formed, and the cosmic microwave background no longer traced the distribution of matter. And this is precisely the most interesting period, when the primordial soup evolved into the rich zoo of objects we now see. *How can astronomers see this dark yet crucial time?*

The situation is similar to having a photo album of a person that begins with the first ultrasound image of him or her as an unborn baby and then skips to some additional photos of his or her years as teenager and adult. The late photos do not simply show a scaled-up version of the first image. We are currently searching for the missing pages of the cosmic photo album that will tell us how the Universe evolved during its infancy to eventually make galaxies like our own Milky Way.

The observers are moving ahead along several fronts. The first involves the construction of large infrared tele-scopes on the ground and in space that will provide us with new (although rather expensive!) photos of galaxies in the Universe at intermediate ages. Current plans include ground-based telescopes which are 24–42 m in diameter, and NASA's successor to the Hubble Space Telescope, the James Webb Space Telescope. In addition, several observational groups around the globe are constructing radio arrays that will be capable of mapping the three-dimensional distribution of cosmic hydrogen left over from the Big Bang in the infant Universe. These arrays are aiming to detect the long-wavelength (redshifted 21-cm) radio emission from hydrogen atoms. Coincidentally, this long wavelength (or low frequency) overlaps with the band used for radio and television broadcasting, and so these telescopes include arrays of regular radio antennas that one can find in electronics stores. These antennas will reveal how the clumpy distribution of neutral hydrogen evolved with cosmic time. By the time the Universe was a few hundreds of millions of years old, the hydrogen distribution had been punched with holes like swiss cheese. These holes were created by the ultraviolet radiation from the first galaxies and black holes, which ionized the cosmic hydrogen in their vicinity.

Theoretical research has focused in recent years on predicting the signals expected from the above instruments and on providing motivation for these ambitious observational projects. In the subsequent chapters of this book, I will describe the theoretical predictions as well as the observational programs planned for testing them. Scientists operate similarly to detectives: they steadily revise their understanding as they collect new information until their model appears consistent with all existing evidence. Their work is exciting as long as it is incomplete.

At a young age I was attracted to philosophy because it addresses the most fundamental questions we face in life. As I matured to an adult, I realized that science has the benefit of formulating a subset of those questions that we can make steady progress on answering, using experimental evidence as a guide.

1.3 Practical Benefits from the Big Picture

I get paid to think about the sky. One might naively regard such an occupation as carrying no practical significance.

If an engineer underestimates the strain on a bridge, the bridge may collapse and harm innocent people. But if I calculate incorrectly the evolution of galaxies, these mistakes bear no immediate consequence for the daily life of other people. *Is this really the case?*

The same engineer who designs bridges would be the first to correct this naive misconception. Newton arrived at his fundamental laws by studying the motion of planets around the Sun, and these laws are now used to build bridges and many other products. Einstein's general theory of relativity was developed to describe the cosmos but is also essential for achieving the required precision in modern navigation or global positioning systems (GPSs) used for both civil and military applications.

But there is a bigger context to the significance of the study of the Universe, namely, cosmology. The big picture gives us the practical advantage of having a more informed view of reality. Consider the weather, for example. It is a natural tendency for people to complain about the harsh weather in particular locations or seasons when rain or snow are common. Some might even associate the weather patterns with a divine entity that reacts to human actions. But if one observes an aerial photo from a satellite, it is easy to understand the origins of the weather patterns in particular locations. The data can be fed into a computer simulation that uses the laws of physics to forecast the weather in advance. With a global understanding of weather and climate patterns one obtains a better sense of reality.

The biggest view we can have is that of the entire Universe. In order to have a balanced worldview we must understand the Universe. When I look up into the dark clear sky at night from the porch of my home in the town of Lexington, Massachusetts, I wonder whether we humans are too often preoccupied with ourselves. There is much more to the Universe than meets the eye around us on Earth.

2

STANDARD COSMOLOGICAL MODEL

2.1 Cosmic Perspective

In 1915 Einstein came up with the general theory of relativity. He was inspired by the fact that all objects follow the same trajectories under the influence of gravity (the so-called equivalence principle, which by now has been tested to better than one part in a trillion), and realized that this would be a natural result if space-time is curved under the influence of matter. He wrote down an equation describing how the distribution of matter (on one side of his equation) determines the curvature of space-time (on the other side of his equation). He then applied his equation to describe the global dynamics of the Universe.

Back in 1915 there were no computers available, and Einstein's equations for the Universe were particularly difficult to solve in the most general case. It was therefore necessary for Einstein to alleviate this difficulty by considering the simplest possible Universe, one that is homogeneous and isotropic. Homogeneity means uniform conditions everywhere (at any given time), and isotropy means the same conditions in all directions when looking out from one vantage point. The combination of these two simplifying assumptions is known as the *cosmological principle*.

The universe can be homogeneous but not isotropic: for example, the expansion rate could vary with direction. It can also be isotropic and not homogeneous: for example, we could be at the center of a spherically symmetric mass distribution. But if it is isotropic around *every* point, then it must also be homogeneous.

Under the simplifying assumptions associated with the cosmological principle, Einstein and his contemporaries were able to solve the equations. They were looking for their "lost keys" (solutions) under the "lamppost" (simplifying assumptions), but the real Universe is not bound by any contract to be the simplest that we can imagine. In fact, it is truly remarkable in the first place that we dare describe the conditions across vast regions of space based on the blueprint of the laws of physics that describe the conditions here on Earth. Our daily life teaches us too often that we fail to appreciate complexity, and that an elegant model for reality is often too idealized for describing the truth (along the lines of approximating a cow as a spherical object).

Back in 1915 Einstein had the wrong notion of the Universe; at the time people associated the Universe with the Milky Way galaxy and regarded all the "nebulae," which we now know are distant galaxies, as constituents within our own Milky Way galaxy. Because the Milky

Way is not expanding, Einstein attempted to reproduce a static universe with his equations. This turned out to be possible after adding a cosmological constant, whose negative gravity would exactly counteract that of matter. However, he soon realized that this solution is unstable: a slight enhancement in density would make the density grow even further. As it turns out, there are no stable static solution to Einstein's equations for a homogenous and isotropic Universe. The Universe must be either expanding or contracting. Less than a decade later, Edwin Hubble discovered that the nebulae previously considered to be constituents of the Milky Way galaxy are receding away from us at a speed v that is proportional to their distance r, namely, $v = H_0 r$ with H_0 a spatial constant (which could evolve with time), commonly termed the *Hubble constant*. Hubble's data indicated that the Universe is expanding.

Einstein was remarkably successful in asserting the cosmological principle. As it turns out, our latest data indicate that the real Universe is homogeneous and isotropic on the largest observable scales to within one part in a hundred thousand. Fortuitously, Einstein's simplifying assumptions turned out to be extremely accurate in describing reality: *the keys were indeed lying next to the lamppost.* Our Universe happens to be the simplest we could have imagined, for which Einstein's equations can be easily solved.

Why was the Universe prepared to be in this special state? Cosmologists were able to go one step further and demonstrate that an early phase transition, called *cosmic inflation*—during which the expansion of the Universe accelerated exponentially—could have naturally produced the conditions postulated by the cosmological principle.

One is left to wonder whether the existence of inflation is just a fortunate consequence of the fundamental laws of nature, or whether perhaps the special conditions of the specific region of space-time we inhabit were selected out of many random possibilities elsewhere by the prerequisite that they allow our existence. The opinions of cosmologists on this question are split.

2.2 Past and Future of Our Universe

Hubble's discovery of the expansion of the Universe has immediate implications with respect to the past and future of the Universe. If we reverse in our mind the expansion history back in time, we realize that the Universe must have been denser in its past. In fact, there must have been a point in time where the matter density was infinite, at the moment of the so-called Big Bang. Indeed we do detect relics from a hotter, denser phase of the Universe in the form of light elements (such as deuterium, helium, and lithium) as well as the cosmic microwave background (CMB). At early times, this radiation coupled extremely well to the cosmic gas and obtained a spectrum known as blackbody, which was predicted a century ago to characterize matter and radiation in equilibrium. The CMB provides the best example of a blackbody spectrum we have.

To get a rough estimate of when the Big Bang occurred, we may simply divide the distance of every galaxy by its recession velocity. This gives a unique answer,

$\sim r/v \sim 1/H_0$, which is independent of distance.* The latest measurements of the Hubble constant give a value of $H_0 \approx 70$ kilometers per second per megaparsec,† implying a current age for the Universe, $1/H_0$, of 14 billion years (or 5×10^{17} seconds).

The second implication concerns our future. A fortunate feature of a spherically symmetric Universe is that, when considering a sphere of matter in it, we are allowed to ignore the gravitational influence of everything outside this sphere. If we empty the sphere and consider a test particle on the boundary of an empty void embedded in a uniform Universe, the particle will experience no net gravitational acceleration. This result, known as Birkhoff's theorem, is reminiscent of Newton's "iron sphere theorem." It allows us to solve the equations of motion for matter on the boundary of the sphere through a local analysis without worrying about the rest of the Universe. Therefore, if the sphere has exactly the same conditions as the rest of the Universe, we may deduce the global expansion history of the Universe by examining its behavior. If the sphere is slightly denser than the mean, we will infer how its density contrast will evolve relative to the background Universe.

The equation describing the motion of a spherical shell of matter is identical to the equation of motion of

---

*Although this is an approximate estimate, it turns out to be within a few percent of the true age of our Universe owing to a coincidence. The cosmic expansion at first decelerated and then accelerated with the two almost canceling each other out at the present time, giving the same age as if the expansion were at a constant speed (as would be strictly true only in an empty Universe).

†A megaparsec (abbreviated as Mpc) is equivalent to 3.086×10^{24} centimeters, or roughly the distance traveled by light in three million years.

a rocket launched from the surface of the Earth. The rocket will escape to infinity if its kinetic energy exceeds its gravitational binding energy, making its total energy positive. However, if its total energy is negative, the rocket will reach a maximum height and then fall back. In order to figure out the future evolution of the Universe, we need to examine the energy of a spherical shell of matter relative to the origin. With a uniform density ρ, a spherical shell of radius r would have a total mass $M = \rho \times (\frac{4\pi}{3}r^3)$ enclosed within it. Its energy per unit mass is the sum of the kinetic energy due to its expansion speed $v = Hr$, $\frac{1}{2}v^2$, and its potential gravitational energy, $-GM/r$ (where G is Newton's constant), namely $E = \frac{1}{2}v^2 - \frac{GM}{r}$. By substituting the above relations for v and M, it can be easily shown that $E = \frac{1}{2}v^2(1 - \Omega)$, where $\Omega = \rho/\rho_c$ and $\rho_c = 3H^2/8\pi G$ is defined as the *critical density*. We therefore find that there are three possible scenarios for the cosmic expansion. The Universe has either (i) $\Omega > 1$, making it gravitationally bound with $E < 0$—*such a "closed Universe" will turn around and end up collapsing toward a "big crunch"*; (ii) $\Omega < 1$, making it gravitationally unbound with $E > 0$—*such an "open Universe" will expand forever*; or the borderline case (iii) $\Omega = 1$, making the Universe marginally bound or "flat" with $E = 0$.

Einstein's equations relate the geometry of space to its matter content through the value of Ω: an open Universe has the geometry of a saddle with a negative spatial curvature, a closed Universe has the geometry of a spherical globe with a positive curvature, and a flat Universe has a flat geometry with no curvature. Our observable section of the Universe appears to be flat.

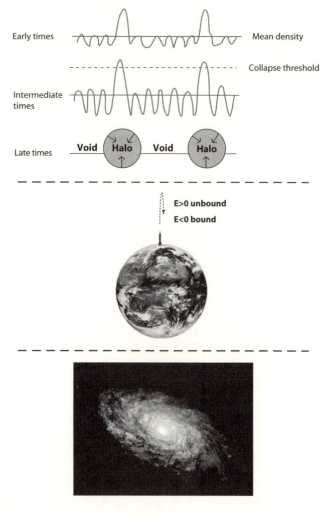

Figure 2.1.

2.3 Gravitational Instability

Now we are at a position to understand how objects, like the Milky Way galaxy, have formed out of small density inhomogeneities that get amplified by gravity.

Let us consider for simplicity the background of a marginally bound (flat) Universe which is dominated by matter. In such a background, only a slight enhancement in density is required to exceed the critical density ρ_c. Because of Birkhoff's theorem, a spherical region that is denser than the mean will behave as if it is part of a closed Universe and increase its density contrast with time, while an underdense spherical region will behave as if it is part of an open Universe and appear more vacant with time relative to the background, as illustrated in figure 2.1.

Figure 2.1. *Facing page, top:* Schematic illustration of the growth of perturbations to collapsed halos through gravitational instability. Once the overdense regions exceed a threshold density contrast above unity, they turn around and collapse to form halos. The material that makes the halos originated in the voids that separate them. *Middle:* A simple model for the collapse of a spherical region. The dynamical fate of a rocket which is launched from the surface of the Earth depends on the sign of its energy per unit mass, $E = v^2/2 - GM_\oplus/r$. The behavior of a spherical shell of matter on the boundary of an overdense region (embedded in a homogeneous and isotropic Universe) can be analyzed in a similar fashion. *Bottom:* A collapsing region may end up as a galaxy, like NGC 4414, shown here (image credit: NASA and ESA). The halo gas cools and condenses to a compact disk surrounded by an extended dark matter halo.

Starting with slight density enhancements that bring them above the critical value ρ_c, the overdense regions will initially expand, reach a maximum radius, and then collapse upon themselves (like the trajectory of a rocket launched straight up, away from the center of the Earth). An initially slightly inhomogeneous Universe will end up clumpy, with collapsed objects forming out of overdense regions. The material to make the objects is drained out of the intervening underdense regions, which end up as voids.

The Universe we live in started with primordial density perturbations of a fractional amplitude $\sim 10^{-5}$. The overdensities were amplified at late times (once matter dominated the cosmic mass budget) up to values close to unity and collapsed to make objects, first on small scales. We have not yet seen the first small galaxies that started the process that eventually led to the formation of big galaxies like the Milky Way. The search for the first galaxies is a search for our origins.

Life as we know it on planet Earth requires water. The water molecule includes oxygen, an element that was not made in the Big Bang and did not exist until the first stars had formed. Therefore our form of life could not have existed in the first hundred millions of years after the Big Bang, before the first stars had formed. There is also no guarantee that life will persist in the distant future.

2.4 Geometry of Space

How can we tell the difference between the flat surface of a book and the curved surface of a balloon? A simple way would be to draw a triangle of straight lines between three points on those surfaces and measure the sum of the three angles of the triangle. The Greek mathematician Euclid demonstrated that the sum of these angles must be 180 degrees (or π radians) on a flat surface. Twenty-one centuries later, the German mathematician Bernhard Riemann extended the field of geometry to curved spaces, which played an important role in the development of Einstein's general theory of relativity. For a triangle drawn on a positively curved surface, like that of a balloon, the sum of the angles is larger than 180 degrees. (This can be easily figured out by examining a globe and noticing that any line connecting one of the poles to the equator opens an angle of 90 degrees relative to the equator. Adding the third angle in any triangle stretched between the pole and the equator would surely result in a total of more than 180 degrees.) According to Einstein's equations, the geometry of the Universe is dictated by its matter content; in particular, the Universe is flat only if the total Ω equals unity. *Is it possible to draw a triangle across the entire Universe and measure its geometry?*

Remarkably, the answer is *yes*. At the end of the twentieth century cosmologists were able to perform this experiment[1] by adopting a simple yardstick provided by the early Universe. The familiar experience of dropping a stone in the middle of a pond results in a circular wave crest that propagates outward. Similarly, perturbation of the smooth Universe at a single point at the Big Bang would have resulted in a spherical sound wave propagating out from that point. The wave would have traveled at the speed of sound, which was of order the speed of light c (or more precisely, $\frac{1}{\sqrt{3}}c$) early on when radiation dominated the cosmic mass budget. At any given time, all the points extending to the distance traveled by the wave are affected by the original pointlike perturbation. The conditions outside this "sound horizon" will remain uncorrelated with the central point, because acoustic information has not been able to reach them at that time. The temperature fluctuations of the CMB trace the simple sum of many such pointlike perturbations that were generated in the Big Bang. The patterns they delineate would therefore show a characteristic correlation scale, corresponding to the sound horizon at the time when the CMB was produced, 400 thousand years after the Big Bang. By measuring the apparent angular scale of this "standard ruler" on the sky, known as the acoustic peak in the CMB, and comparing it to theory, experimental cosmologists inferred from the simple geometry of triangles that the Universe is flat.

The inferred flatness is a natural consequence of the early period of vast expansion, known as cosmic inflation, during which any initial curvature was flattened. Indeed a small patch of a fixed size (representing our current observable region in the cosmological context) on the surface of a vastly inflated balloon would appear nearly flat. The sum of the angles on a nonexpanding triangle placed on this patch will get arbitrarily close to 180 degrees as the balloon inflates.

2.5 Cosmic Archaeology

Our Universe is the simplest possible on two counts: it satisfies the cosmological principle, and it has a flat geometry. The mathematical description of an expanding, homogeneous, and isotropic Universe with a flat geometry is straightforward. We can imagine filling up space with clocks that are all synchronized. At any given snapshot in time the physical conditions (density, temperature) are the same everywhere. But as time goes on, the spatial separation between the clocks will increase. The stretching of space can be described by a time-dependent scale factor $a(t)$. A separation measured at time t_1 as $r(t_1)$ will appear at time t_2 to have a length $r(t_2) = r(t_1)[a(t_2)/a(t_1)]$.

A natural question to ask is whether our human bodies, or even the solar system, are also expanding as the Universe expands. The answer is no, because these systems are held together by forces whose strength far exceeds the cosmic force. The mean density of the Universe today, $\bar{\rho}$, is 29 orders of magnitude smaller than the density of our bodies. Not only are the electromagnetic forces that keep the atoms in our body together far greater than gravity, but even the gravitational self-force of our body on itself overwhelms the cosmic influence. Only on very large scales does the cosmic gravitational force dominate the scene. This also implies that we cannot observe the cosmic expansion with a local laboratory experiment; in order to notice the expansion we need to observe sources that are spread over the vast scales of millions of light years.

A source located at a separation $r = a(t)x$ from us would move at a velocity $v = dr/dt = \dot{a}x = (\dot{a}/a)r$, where $\dot{a} = da/dt$. Here x is a time-independent tag, denoting the present-day distance of the source. Defining $H = \dot{a}/a$ which is constant in space, we recover the Hubble expansion law $v = Hr$.

Edwin Hubble measured the expansion of the Universe using the Doppler effect. We are all familiar with the same effect for sound waves: when a moving car sounds its horn, the pitch (frequency) we hear is different if the car is approaching us or moving away. Similarly, the wavelength of light depends on the velocity of the source relative to us. As the Universe expands, a light source will move away from us and its Doppler effect will change with time. The Doppler formula for a near-by source of light (with a recession speed much smaller than the speed of light) gives $\Delta v / v \approx -\Delta v / c = -(\dot{a}/a)(r/c) = -(\dot{a}\,\Delta t)/a = -\Delta a / a$, admitting the solution $v \propto a^{-1}$. Correspondingly, the wavelength scales as $\lambda = (c/v) \propto a$. We could have anticipated this outcome since a wavelength can be used as a measure of distance and should therefore be stretched as the Universe expands. The redshift z is defined through the factor $(1 + z)$ by which the photon wavelength was stretched (or its frequency reduced) between its emission and observation times. If we define $a = 1$ today, then $a = 1/(1 + z)$ at earlier times. Higher redshifts correspond to a higher recession speed of the source relative to us (ultimately approaching the speed of light when the redshift goes to infinity), which in turn implies a larger distance (ultimately approaching our horizon, which is the distance traveled by light since the Big Bang) and an earlier emission time of the source in order for the photons to reach us today.

We see high-redshift sources as they looked at early cosmic times. Observational cosmology is like archaeology—the deeper we look into space, the more ancient are the clues about our history (see figure 2.2). But there is a limit

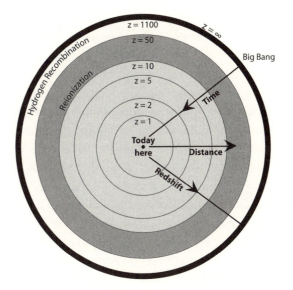

Figure 2.2. Cosmic archaeology of the observable volume of the Universe, in comoving coordinates (which factor out the cosmic expansion). The outermost observable boundary ($z = \infty$) marks the comoving distance that light has traveled since the Big Bang. Future observatories aim to map most of the observable volume of our Universe and dramatically improve the statistical information we have about the density fluctuations within it. Existing data on the CMB probe mainly a very thin shell at the hydrogen recombination epoch ($z \sim 10^3$, beyond which the Universe is opaque), and current large-scale galaxy surveys map only a small region near us at the center of the diagram. The formation epoch of the first galaxies that culminated with hydrogen reionization at a redshift $z \sim 10$ is shaded gray. Note that the comoving volume out to any of these redshifts scales as the distance cubed.

to how far back we can see. In principle, we can image the Universe only as long as it was transparent, corresponding to redshifts $z < 10^3$ for photons. The first galaxies are believed to have formed long after that.

The expansion history of the Universe is captured by the scale factor $a(t)$. We can write a simple equation for the evolution of $a(t)$ based on the behavior of a small region of space. For that purpose we need to incorporate the fact that in Einstein's theory of gravity, not only does mass density ρ gravitate, but pressure p does too. In a homogeneous and isotropic Universe, the quantity $\rho_{\text{grav}} = (\rho + 3p/c^2)$ plays the role of the gravitating mass density ρ of Newtonian gravity.[2] There are several examples to consider. For a radiation fluid,* $p_{\text{rad}}/c^2 = \frac{1}{3}\rho_{\text{rad}}$, implying that $\rho_{\text{grav}} = 2\rho_{\text{rad}}$. On the other hand, for a constant vacuum density (the so-called "cosmological constant"), the pressure is negative because by opening up a new volume increment ΔV one gains an energy $\rho c^2 \Delta V$ instead of losing energy, as is the case for normal fluids that expand into more space. In thermodynamics, pressure is derived from the deficit in energy per unit of new volume, which in this case gives $p_{\text{vac}}/c^2 = -\rho_{\text{vac}}$. This in turn leads to another reversal of signs, $\rho_{\text{grav}} = (\rho_{\text{vac}} + 3p_{\text{vac}}/c^2) = -2\rho_{\text{vac}}$, which may be interpreted as repulsive gravity! This surprising result gives rise to the phenomenon of accelerated cosmic expansion, which characterized the early period of cosmic inflation as well as the latest six billion years of cosmic history.

---

*The momentum of each photon is $1/c$ of its energy. The pressure is defined as the momentum flux along one dimension out of three, and is therefore given by $\rho_{\text{rad}}c^2/3$, where ρ_{rad} is the mass density of the radiation.

As the Universe expands and the scale factor increases, the matter mass density declines inversely with volume, $\rho_{matter} \propto a^{-3}$, whereas the radiation energy density decreases as $\rho_{rad} c^2 \propto a^{-4}$, because not only is the density of photons diluted as a^{-3}, but the energy per photon $h\nu = hc/\lambda$ (where h is Planck's constant) declines as a^{-1}. Today ρ_{matter} is larger than ρ_{rad} by a factor of $\sim 3{,}300$, but at $(1 + z) \sim 3{,}300$ the two were equal, and at even higher redshifts the radiation dominated. Since a stable vacuum does not get diluted with cosmic expansion, the present-day ρ_{vac} remained a constant and dominated over ρ_{matter} and ρ_{rad} only at late times (whereas the unstable "false vacuum" that dominated during inflation had decayed when inflation ended).

2.6 Milestones in Cosmic Evolution

The gravitating mass, $M_{grav} = \rho_{grav} V$, enclosed by a spherical shell of radius $a(t)$ and volume $V = \frac{4\pi}{3} a^3$, induces an acceleration

$$\frac{d^2 a}{d t^2} = -\frac{G M_{grav}}{a^2}. \qquad (2.1)$$

Since $\rho_{grav} = \rho + 3p/c^2$, we need to know how pressure evolves with the expansion factor $a(t)$. This is obtained from the thermodynamic relation mentioned above between the change in the internal energy $d(\rho c^2 V)$ and the $p\,dV$ work done by the pressure, $d(\rho c^2 V) = -p\,dV$. This relation implies $-3p a \dot{a}/c^2 = a^2 \dot{\rho} + 3\rho a \dot{a}$, where an overdot denotes a time derivative. Multiplying equation (2.1) by $\dot{a}$ and making use of this relation yields our familiar result

$$E = \frac{1}{2}\dot{a}^2 - \frac{GM}{a}, \qquad (2.2)$$

where E is a constant of integration and $M \equiv \rho V$. As discussed before, the spherical shell will expand forever (being gravitationally unbound) if $E \geq 0$, but will eventually collapse (being gravitationally bound) if $E < 0$. By making use of the Hubble parameter, $H = \dot{a}/a$, equation (2.2) can be rewritten as

$$\frac{E}{\frac{1}{2}\dot{a}^2} = 1 - \Omega, \qquad (2.3)$$

where $\Omega = \rho/\rho_c$, with

$$\rho_c = \frac{3H^2}{8\pi G} = \left(9.2 \times 10^{-30} \frac{\text{g}}{\text{cm}^3}\right)\left(\frac{H}{70\,\text{km s}^{-1}\text{Mpc}^{-1}}\right)^2. \qquad (2.4)$$

With Ω_m, Ω_Λ, and Ω_r denoting the present contributions to Ω from matter (including cold dark matter [see section 2.7] as well as a contribution Ω_b from ordinary matter of protons and neutrons, or "baryons"), vacuum density (cosmological constant), and radiation, respectively, a flat universe satisfies

$$\frac{H(t)}{H_0} = \left(\frac{\Omega_m}{a^3} + \Omega_\Lambda + \frac{\Omega_r}{a^4} \right)^{1/2}, \qquad (2.5)$$

where we define H_0 and $\Omega_0 = \Omega_m + \Omega_\Lambda + \Omega_r = 1$ to be the present-day values of H and Ω, respectively.

In the particularly simple case of a flat Universe, we find that if matter dominates then $a \propto t^{2/3}$, if radiation dominates then $a \propto t^{1/2}$, and if the vacuum density dominates then $a \propto \exp\{H_{vac}t\}$ with $H_{vac} = (8\pi G\rho_{vac}/3)^{1/2}$ being a constant. In the beginning, after inflation ended, the mass density of our Universe ρ was at first dominated by radiation at redshifts $z > 3{,}300$, then it became dominated by matter at $0.3 < z < 3{,}300$, and finally was dominated by the vacuum at $z < 0.3$. The vacuum started to dominate ρ_{grav} already at $z < 0.7$, or six billion years ago. Figure 2.4 below illustrates the mass budget in the present-day Universe and during the epoch when the first galaxies had formed.

The above results for $a(t)$ have two interesting implications. First, we can figure out the relationship between the time since the Big Bang and redshift since $a = (1 + z)^{-1}$. For example, during the matter-dominated era ($1 < z < 10^3$),

$$t \approx \frac{2}{3H_0\Omega_m^{1/2}(1 + z)^{3/2}} = \frac{0.95 \times 10^9 \, \text{yr}}{[(1 + z)/7]^{3/2}}. \qquad (2.6)$$

Second, we note the remarkable exponential expansion for a vacuum-dominated phase. This accelerated expansion serves an important purpose in explaining a few puzzling features of our Universe. We already noticed that our Universe was prepared in a very special initial state: nearly isotropic and homogeneous, with Ω close to unity and a flat geometry. In fact, it took the CMB photons nearly the entire age of the Universe to travel toward us. Therefore, it should take them twice as long to bridge their points of origin on opposite sides of the sky. *How is it possible then that the conditions of the Universe (as reflected in the nearly uniform CMB temperature) were prepared to be the same in regions that were never in causal contact before?* Such a degree of organization is highly unlikely to occur at random. If we receive our clothes ironed out and folded neatly, we know that there must have been a process that caused it. Cosmologists have identified an analogous "ironing process" in the form of *cosmic inflation*. This process is associated with an early period during which the Universe was dominated temporarily by the mass density of an elevated vacuum state, and experienced exponential expansion by at least ~ 60 e-folds. This vast expansion "ironed out" any initial curvature of our environment, and generated a flat geometry and nearly uniform conditions across a region far greater than our current horizon. After the elevated vacuum state decayed, the Universe became dominated by radiation.

The early epoch of inflation is important not just in producing the global properties of the Universe but also in generating the inhomogeneities that seeded the formation of galaxies within it.[3] The vacuum energy density that had driven inflation encountered quantum-mechanical fluctuations. After the perturbations were stretched beyond the horizon of the infant Universe (which today would have occupied the size no bigger than a human hand), they materialized as perturbations in the mass density of radiation and matter. The last perturbations to leave the horizon during inflation eventually entered back after inflation ended (when the scale factor grew more slowly than ct). It is tantalizing to contemplate the notion that galaxies, which represent massive classical objects with $\sim 10^{67}$ atoms in today's Universe, might have originated from subatomic quantum-mechanical fluctuations at early times.

After inflation, an unknown process, called "baryogenesis" or "leptogenesis," generated an excess of particles (baryons and leptons) over antiparticles.* As the Universe cooled to a temperature of hundreds of MeV (with $1\,\mathrm{MeV}/k_B = 1.1604 \times 10^{10}\,\mathrm{K}$), protons and neutrons condensed out of the primordial quark-gluon plasma through the so-called *QCD phase transition*. At about one second after the Big Bang, the temperature declined to ~ 1 MeV, and the weakly interacting neutrinos decoupled. Shortly afterward the abundance of neutrons relative to protons froze and electrons and positrons annihilated. In the next few minutes, nuclear fusion reactions produced light elements more massive than hydrogen, such as deuterium, helium, and lithium, in abundances that match those observed today in regions where gas has not been

---

*Antiparticles are identical to particles but with opposite electric charge. Today, the ordinary matter in the Universe is observed to consist almost entirely of particles. The origin of the asymmetry in the cosmic abundance of matter over antimatter is still an unresolved puzzle.

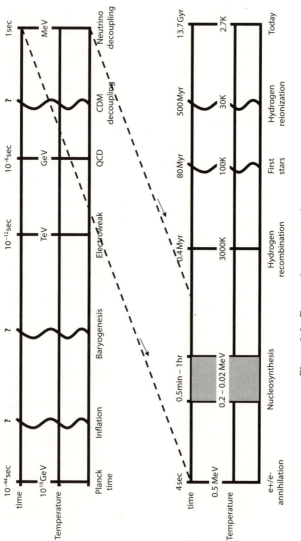

Figure 2.3. For caption see opposite page.

processed subsequently through stellar interiors. Although the transition to matter domination occurred at a redshift $z \sim 3,300$, the Universe remained hot enough for the gas to be ionized, and electron-photon scattering effectively coupled ordinary matter and radiation. At $z \sim 1,100$ the temperature dipped below $\sim 3,000$ K, and free electrons recombined with protons to form neutral hydrogen atoms. As soon as the dense fog of free electrons was depleted, the Universe became transparent to the relic radiation, which is observed at present as the CMB. These milestones of the thermal history are depicted in figure 2.3.

The Big Bang is the only known event in our past history where particles interacted with center-of-mass energies

---

Figure 2.3. Following inflation, the Universe went through several other milestones which left a detectable record. These include baryogenesis (which resulted in the observed asymmetry between matter and antimatter), the electroweak phase transition (during which the symmetry between electromagnetic and weak interactions was broken), the QCD phase transition (during which protons and neutrons nucleated out of a soup of quarks and gluons), the dark matter decoupling epoch (during which the dark matter decoupled thermally from the cosmic plasma), neutrino decoupling, electron-positron annihilation, light element nucleosynthesis (during which helium, deuterium, and lithium were synthesized), and hydrogen recombination. The cosmic time and CMB temperature of the various milestones are marked. Wavy lines and question marks indicate milestones with uncertain properties. The signatures that the same milestones left in the Universe are used to constrain its parameters.

approaching the so-called Planck scale* $[(hc^5/G)^{1/2} \sim 10^{19}\,\text{GeV}]$, at which quantum mechanics and gravity are expected to be unified. Unfortunately, the exponential expansion of the Universe during inflation erased memory of earlier cosmic epochs, such as the Planck time.

2.7 Most Matter Is Dark

Surprisingly, most of the matter in the Universe is not the same ordinary matter that we are made of (see figure 2.4). If it were ordinary matter (which also makes stars and diffuse gas), it would have interacted with light, thereby revealing its existence to observations through telescopes. Instead, observations of many different astrophysical environments require the existence of some mysterious dark component of matter which reveals itself only through its gravitational influence and leaves no other clue about its nature. Cosmologists are like a detective who finds evidence for some unknown criminal in a crime scene and is anxious to find his/her identity. The evidence for dark matter is clear and indisputable, assuming that the laws of gravity are not modified (although a small minority of scientists are exploring this alternative).

Without dark matter we would not have come into existence by now. This is because ordinary matter is coupled to the CMB radiation that filled up the Universe early

*The Planck energy scale is obtained by equating the quantum-mechanical wavelength of a relativistic particle with energy E, namely, hc/E, to its "black hole" radius $\sim GE/c^4$, and solving for E.

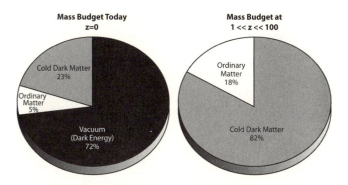

Figure 2.4. Mass budgets of different components in the present day Universe and in the infant Universe when the first galaxies formed (redshifts $z = 10$–50). The CMB radiation (not shown) makes up a fraction $\sim$0.005% of the budget today, but was dominant at redshifts $z \gg 3{,}300$. The cosmological constant (vacuum) contribution was negligible at high redshifts ($z \gg 1$).

on. The diffusion of photons on small scales smoothed out perturbations in this primordial radiation fluid. The smoothing length was stretched to a scale as large as hundreds of millions of light years in the present-day Universe. This is a huge scale by local standards, since galaxies—like the Milky Way—were assembled out of matter in regions a hundred times smaller than that. Because ordinary matter was coupled strongly to the radiation in the early dense phase of the Universe, it also was smoothed on small scales. If there was nothing else in addition to the radiation and ordinary matter, then this smoothing process would have had a devastating effect on the prospects for life in our Universe. Galaxies like the Milky Way would not have formed by the present time since there would have been no density perturbations on the relevant small scales to seed their formation. The existence of dark matter not coupled to the radiation came to the rescue by keeping memory of the initial seeds of density perturbations on small scales. In our neighborhood, these seed perturbations led eventually to the formation of the Milky Way galaxy inside of which the Sun was made as one out of tens of billions of stars, and the Earth was born out of the debris left over from the formation process of the Sun. This sequence of events would never have occurred without the dark matter.

We do not know what the dark matter is made of, but from the good match obtained between observations of large-scale structure and the equations describing a pressureless fluid [see equations (3.3) and (3.4)], we infer that it is likely made of particles with small random velocities. It is therefore called "cold dark matter" (CDM). The popular view is that CDM is composed of particles that possess weak interactions with ordinary matter, similarly to the elusive neutrinos we know to exist. The hope is that CDM particles, owing to their weak but nonvanishing coupling to ordinary matter, will nevertheless be produced in small quantities through collisions of energetic particles in future laboratory experiments such as the Large Hadron Collider (LHC).[4] Other experiments are attempting to detect directly the astrophysical CDM particles in the Milky Way halo. A positive result from any of these experiments will be equivalent to our detective friend being successful in finding a DNA sample of the previously unidentified criminal.

According to the standard cosmological model, the CDM behaves as a collection of collisionless particles that started out at the epoch of matter domination with negligible thermal velocities, and later evolved exclusively under gravitational forces. The model explains how both individual galaxies and the large-scale patterns in their distribution originated from the small, initial density fluctuations. On the largest scales, observations of the present galaxy distribution have indeed found the same statistical patterns as seen in the CMB, enhanced as expected by billions of years of gravitational evolution. On smaller scales, the model describes how regions that were denser than average collapsed due to their enhanced gravity and eventually formed gravitationally bound halos, first on small spatial scales and later on larger ones. In this hierarchical model of galaxy formation, the small galaxies formed first and then merged, or accreted gas, to form larger galaxies. At each snapshot of this cosmic evolution, the abundance of collapsed halos, whose masses are dominated by dark matter, can be computed from the initial conditions. The common understanding of galaxy formation is based on the notion that stars formed out of the gas that cooled and subsequently condensed to high densities in the cores of some of these halos.

Gravity thus explains how some gas is pulled into the deep potential wells within dark matter halos and forms galaxies. One might naively expect that the gas outside halos would remain mostly undisturbed. However, observations show that it has not remained neutral (i.e., in atomic form), but was largely ionized by the UV radiation emitted by the galaxies. The diffuse gas pervading the space outside and between galaxies is referred to as the intergalactic medium (IGM). For the first hundreds of millions of years after cosmological recombination (when protons and electrons combined to make neutral hydrogen), the so-called cosmic dark ages, the universe was filled with diffuse atomic hydrogen. As soon as galaxies formed, they started to ionize diffuse hydrogen in their vicinity. Within less than a billion years, most of the IGM was reionized.

3

THE FIRST GAS CLOUDS

The initial conditions of the Universe can be summarized on a single sheet of paper. The small number of parameters that provide an accurate statistical description of these initial conditions are summarized in table 3.1. However, thousands of books in libraries throughout the world cannot summarize the complexities of galaxies, stars, planets, life, and intelligent life in the present-day Universe. If we feed the simple initial cosmic conditions into a gigantic computer simulation incorporating the known laws of physics, we should be able to reproduce all the complexity that emerged out of the simple early Universe. Hence, all the information associated with this later complexity was encapsulated in those simple initial conditions. Below we follow the process through which late time complexity appeared and established an irreversible arrow to the flow of cosmic time.*

*In previous decades, astronomers used to associate the simplicity of the early Universe with the fact that the data about it were scarce. Although this was true at the infancy of observational cosmology, it is not true any more. With much

TABLE 3.1

Standard set of cosmological parameters (defined and adopted throughout the book).

Ω_Λ	Ω_m	Ω_b	h_0	n_s	σ_8
0.72	0.28	0.05	0.7	1	0.82

Source: E. Komatsu et al., *Astrophys. J. Suppl.* **180**, 330 (2009).

After cosmological recombination, the Universe entered the "dark ages" during which the relic CMB light from the Big Bang gradually faded away. During this "pregnancy" period, which lasted hundreds of millions of years, the seeds of small density fluctuations planted by inflation in the matter distribution grew up until they eventually collapsed to make the first galaxies.[5]

3.1 Growing the Seed Fluctuations

As discussed earlier, small perturbations in density grow due to the unstable nature of gravity. Overdense regions behave as if they reside in a closed Universe. Their evolution ends in a "big crunch," which results in the formation of gravitationally bound objects like the Milky Way galaxy.

Equation (2.3) explains the formation of galaxies out of seed density fluctuations in the early Universe, at a time when the mean matter density was very close to the critical value and $\Omega_m \approx 1$. Given that the mean cosmic density

---

richer data in our hands, the initial simplicity is now interpreted as an outcome of inflation.

was close to the threshold for collapse, a spherical region which was only slightly denser than the mean behaved as if it was part of an $\Omega > 1$ Universe, and therefore eventually collapsed to make a bound object, like a galaxy. The material from which objects are made originated in the underdense regions (voids) that separate these objects (and which behaved as part of an $\Omega < 1$ Universe), as illustrated in figure 2.1.

Observations of the CMB show that at the time of hydrogen recombination the Universe was extremely uniform, with spatial fluctuations in the energy density and gravitational potential of roughly one part in 10^5. These small fluctuations grew over time during the matter-dominated era as a result of gravitational instability, and eventually led to the formation of galaxies and larger-scale structures, as observed today.

In describing the gravitational growth of perturbations in the matter-dominated era ($z \ll 3,300$), we may consider small perturbations of a fractional amplitude $|\delta| \ll 1$ on top of the uniform background density $\bar{\rho}$ of cold dark matter. The three fundamental equations describing conservation of mass and momentum along with the gravitational potential can then be expanded to leading order in the perturbation amplitude. We distinguish between physical and comoving coordinates (the latter expanding with the background Universe). Using vector notation, the fixed coordinate $\mathbf{r}$ corresponds to a comoving position $\mathbf{x} = \mathbf{r}/a$. We describe the cosmological expansion in terms of an ideal pressureless fluid of particles, each of which is at fixed $\mathbf{x}$, expanding with the Hubble flow $\mathbf{v} = H(t)\mathbf{r}$, where $\mathbf{v} = d\mathbf{r}/dt$. Onto this uniform expansion we impose small fractional density perturbations

$$\delta(\mathbf{x}) = \frac{\rho(\mathbf{r})}{\bar{\rho}} - 1, \qquad (3.1)$$

where the mean fluid mass density is $\bar{\rho}$, with a corresponding peculiar velocity which describes the deviation from the Hubble flow $\mathbf{u} \equiv \mathbf{v} - H\mathbf{r}$. The fluid is then described by the continuity and Euler equations in comoving coordinates:

$$\frac{\partial \delta}{\partial t} + \frac{1}{a} \nabla \cdot [(1 + \delta)\mathbf{u}] = 0, \qquad (3.2)$$

$$\frac{\partial \mathbf{u}}{\partial t} + H\mathbf{u} + \frac{1}{a}(\mathbf{u} \cdot \nabla)\mathbf{u} = -\frac{1}{a}\nabla\phi. \qquad (3.3)$$

The gravitational potential ϕ is given by the Newtonian Poisson equation, in terms of the density perturbation:

$$\nabla^2 \phi = 4\pi G \bar{\rho} a^2 \delta. \qquad (3.4)$$

This fluid description is valid for describing the evolution of collisionless cold dark matter particles until different particle streams cross. The crossing typically occurs only after perturbations have grown to become nonlinear with $|\delta| > 1$, and at that point the individual particle trajectories must in general be followed.

The combination of the above equations yields to leading order in δ

$$\frac{\partial^2 \delta}{\partial t^2} + 2H\frac{\partial \delta}{\partial t} = 4\pi G \bar{\rho}\delta. \qquad (3.5)$$

This linear equation has in general two independent solutions, only one of which grows in time. Starting with random initial conditions, this "growing mode" comes to dominate the density evolution. Thus, until it becomes nonlinear, the density perturbation maintains its shape in comoving coordinates and grows in amplitude in proportion to a growth factor $D(t)$. The growth factor in a flat Universe at $z < 10^3$ is given by*

$$D(t) \propto \frac{\left(\Omega_\Lambda a^3 + \Omega_m\right)^{1/2}}{a^{3/2}} \int_0^a \frac{a'^{3/2}\, da'}{\left(\Omega_\Lambda a'^3 + \Omega_m\right)^{3/2}}. \quad (3.6)$$

In the matter-dominated regime of the redshift range $1 < z < 10^3$, the growth factor is simply proportional to the scale factor $a(t)$. Interestingly, the gravitational potential $\phi \propto \delta/a$ does not grow in comoving coordinates. This implies that the potential depth fluctuations remain frozen in amplitude as fossil relics from the inflationary epoch during which they were generated. Nonlinear collapse only changes the potential depth by a factor of order unity, but even inside collapsed objects its rough magnitude remains as testimony to the inflationary conditions. This explains why the characteristic potential depth of collapsed objects such as galaxy clusters ($\phi/c^2 \sim 10^{-5}$) is of the same order as the potential fluctuations probed by the fractional variations in the CMB temperature across the sky. At low redshifts $z < 1$ and in the future, the cosmological constant dominates ($\Omega_m \ll \Omega_\Lambda$) and the density fluctuations

---

*An analytic expression for the growth factor in terms of special functions was derived by D. Eisenstein, http://arxiv.org/pdf/astro-ph/9709054v2 (1997).

freeze in amplitude [$D(t) \rightarrow$ constant] as their growth is suppressed by the accelerated expansion of space.

The initial perturbation amplitude varies with spatial scale. Large-scale regions have a smaller perturbation amplitude than small-scale regions. The statistical properties of the perturbations as a function of spatial scale can be captured by expressing the density field as a sum over a complete set of periodic "Fourier modes," each having a sinusoidal (wavelike) dependence on space with a comoving wavelength $\lambda = 2\pi/k$ and wavenumber k. Mathematically, we write $\delta_{\mathbf{k}} = \int d^3x \, \delta(x) e^{-i\mathbf{k}\cdot\mathbf{x}}$, with $\mathbf{x}$ being the comoving spatial coordinate. The characteristic amplitude of each $\mathbf{k}$-mode defines the typical value of δ on the spatial scale λ. Inflation generates perturbations in which different $\mathbf{k}$-modes are statistically independent, and each has a random phase constant in its sinusoid. The statistical properties of the fluctuations are determined by the variance of the different $\mathbf{k}$-modes given by the so-called power spectrum, $P(k) = (2\pi)^{-3} \langle |\delta_{\mathbf{k}}|^2 \rangle$, where the angular brackets denote an average over the entire statistical ensemble of modes.

In the standard cosmological model, inflation produces a primordial power-law spectrum $P(k) \propto k^{n_s}$ with $n_s \approx 1$. This spectrum admits the special property that gravitational potential fluctuations of all wavelengths have the same amplitude at the time when they enter the horizon (namely, when their wavelength matches the distance traveled by light during the age of the Universe), and so this spectrum is called "scale invariant." The growth of perturbations in a CDM Universe results in a modified final power spectrum characterized by a turnover at a scale of order the horizon cH^{-1} at matter-radiation equality, and a small-scale asymptotic shape of $P(k) \propto k^{n_s-4}$. The break is generated by the fact that there was little growth of the perturbations when the Universe was dominated by radiation. The overall amplitude of the power spectrum is not specified by current models of inflation, and is usually set by comparing to the observed CMB temperature fluctuations or to measures of large-scale structure based on surveys of galaxies or the intergalactic gas (the so-called "Lyman-α forest," to be discussed in chapter 7).

In order to determine the formation of objects of a given size or mass it is useful to consider the statistical distribution of the smoothed density field. To smooth the density distribution, cosmologists use a window (or filter) function $W(\mathbf{r})$ normalized so that $\int d^3r\, W(\mathbf{r}) = 1$, with the smoothed density perturbation field being $\int d^3r\, \delta(\mathbf{x})\, W(\mathbf{r})$. For the particular choice of a spherical top-hat window (similar to a cookie cutter), in which $W = 1$ in a sphere of radius R and $W = 0$ outside the sphere, the smoothed perturbation field measures the fluctuations in the mass in spheres of radius R. The normalization of the present power spectrum at $z = 0$ is often specified by the value of $\sigma_8 \equiv \sigma(R = 8h_0^{-1}\mathrm{Mpc})$ where $h_0 = 0.7$ calibrates the Hubble constant today as $H_0 = 100h_0$ km s^{-1} Mpc^{-1}. For the top-hat filter, the smoothed perturbation field is denoted by δ_R or δ_M, where the enclosed mass M is related to the comoving radius R by $M = 4\pi\rho_m R^3/3$, in terms of current mean density of matter ρ_m. The variance $\langle \delta_M^2 \rangle$ is

$$\sigma^2(M) \equiv \sigma^2(R) = \int_0^\infty \frac{dk}{2\pi^2}\, k^2 P(k) \left(\frac{3j_1(kR)}{kR} \right)^2,$$

$$(3.7)$$

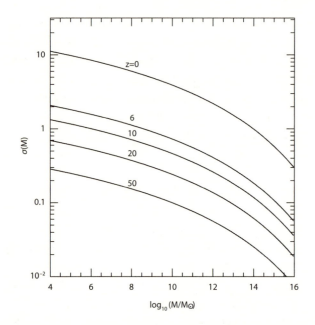

Figure 3.1. The root-mean-square amplitude of linearly extrapolated density fluctuations σ as a function of mass M (in solar masses
$M_\odot$, within a spherical top-hat filter) at different redshifts z. Halos
form in regions that exceed the background density by a factor of
order unity. This threshold is surpassed only by rare (many-σ) peaks
for high masses at high redshifts. When discussing the abundance
of halos in section 3.4, we will factor out the linear growth of
perturbations and use the function $\sigma(M)$ at $z=0$.

where $j_1(x) = (\sin x - x \cos x)/x^2$. The term involving
j_1 in the integrand is the Fourier transform of $W(\mathbf{r})$.
The function $\sigma(M)$ plays a crucial role in estimates of
the abundance of collapsed objects, and is plotted in
figure 3.1 as a function of mass and redshift for the

standard cosmological model. For modes with random phases, the probability of different regions with the same size to have a perturbation amplitude between δ and $\delta + d\delta$ is Gaussian with a zero mean and the above variance, $P(\delta)d\delta = (2\pi\sigma^2)^{-1/2}\exp\{-\delta^2/2\sigma^2\}d\delta$.

3.2 The Smallest Gas Condensations

As the density contrast between a spherical gas cloud and its cosmic environment grows, there are two main forces which come into play. The first is *gravity* and the second is *pressure*. We can get a rough estimate for the relative importance of these forces from the following simple considerations. The increase in gas density near the center of the cloud sends out a pressure wave which propagates out at the speed of sound $c_s \sim (kT/m_p)^{1/2}$ where T is the gas temperature. The wave tries to even out the density enhancement, consistent with the tendency of pressure to resist collapse. At the same time, gravity pulls the cloud together in the opposite direction. The characteristic time scale for the collapse of the cloud is given by its radius R divided by the free-fall speed $\sim (2GM/R)^{1/2}$, yielding $t_{\text{coll}} \sim (G\langle\rho\rangle)^{-1/2}$ where $\langle\rho\rangle = M/\frac{4\pi}{3}R^3$ is the characteristic density of the cloud as it turns around on its way to collapse.* If the sound wave does not have sufficient

---

*Substitution of the mean density of the Earth into this expression yields the characteristic time it takes a freely falling elevator to reach the center of the Earth from its surface ($\sim 1/3$ of an hour), as well as the order of magnitude of the time it takes a low-orbit satellite to go around the Earth (~ 1.5 hours).

time to traverse the cloud during the free-fall time, namely, $R > R_J \equiv c_s t_{coll}$, then the cloud will collapse. Under these circumstances, the sound wave moves outward at a speed that is slower than the inward motion of the gas, and so the wave is simply carried along together with the infalling material. On the other hand, the collapse will be inhibited by pressure for a sufficiently small cloud with $R < R_J$. The transition between these regimes is defined by the so-called Jeans radius R_J, corresponding to the Jeans mass

$$M_J = \frac{4\pi}{3} \langle\rho\rangle R_J^3. \qquad (3.8)$$

This mass corresponds to the total gravitating mass of the cloud, including the dark matter. As long as the gas temperature is not very different from the CMB temperature, the value of $M_J \sim 10^5 M_\odot$ is independent of redshift.[6] This is the minimum total mass of the first gas cloud to collapse ~100 million years after the Big Bang. A few hundred million years later, once the cosmic gas was ionized and heated to a temperature $T > 10^4$ K by the first galaxies, the minimum galaxy mass had risen above ~$10^8 M_\odot$. At even later times, the UV light that filled up the Universe was able to boil the uncooled gas out of the shallowest gravitational potential wells of minihalos with a characteristic temperature below 10^4 K.[7]

As mentioned in chapter 2, existing cosmological data suggest that the dark matter is "cold," that is, its pressure is negligible during the gravitational growth of galaxies. In popular models, the Jeans mass of the dark matter alone is negligible but nonzero, of the order of the mass of a planet like Earth or Jupiter.[8] All halos between this minimum clump mass and $\sim 10^5 M_\odot$ are expected to contain mostly dark matter and little ordinary matter.

3.3 Spherical Collapse and Halo Properties

When an object above the Jeans mass collapses, the dark matter forms a halo inside which the gas may cool, condense to the center, and eventually fragment into stars. The dark matter cannot cool since it has very weak interactions. As a result, a galaxy emerges with a central core that is occupied by stars and cold gas and is surrounded by an extended halo of invisible dark matter. Since cooling eliminates the pressure support from the gas, the only force that can prevent the gas from sinking all the way to the center and ending up in a black hole is the centrifugal force associated with its rotation around the center (angular momentum). The slight ($\sim 5\%$) rotation, given to the gas by tidal torques from nearby galaxies as it turns around from the initial cosmic expansion and gets assembled into the object, is sufficient to stop its infall on a scale which is *an order of magnitude smaller* than the size of the dark matter halo[9] (the so-called "virial radius"). On this stopping scale, the gas is assembled into a thin disk and orbits around the center for an extended period of time, during which it tends to break into dense clouds which fragment further into denser clumps. Within the compact clumps that are produced, the gas density is sufficiently high and the gas temperature is sufficiently low for the Jeans mass to be of order the mass of a star. As a result, the clumps fragment into stars and a galaxy is born.

In the popular cosmological model, small objects formed first. The very first stars must have therefore formed inside gas condensations just above the cosmological Jeans mass, $\sim 10^5 M_{\odot}$. Whereas each of these first gaseous halos was not massive or cold enough to make more than a single high-mass star, star clusters started to form shortly afterward inside bigger halos. By solving the equation of motion (2.1) for a spherical overdense region, it is possible to relate the characteristic radius and gravitational potential well of each of these galaxies to their mass and their redshift of formation.

The small density fluctuations evidenced in the CMB grew over time as described in section 3.1, until the perturbations δ became of order unity and the full nonlinear gravitational collapse followed. The dynamical collapse of a dark matter halo can be solved analytically in spherical symmetry with an initial top-hat region of uniform overdensity δ_i inside a sphere of radius R. Although this toy model might seem artificially simple, its results have turned out to be surprisingly accurate for interpreting the properties and distribution of halos in numerical simulations of cold dark matter.

During the gravitational collapse of a spherical region, the enclosed overdensity δ grows initially as $\delta_L = \delta_i D(t)/D(t_i)$, in accordance with linear theory, but eventually δ grows above δ_L. Any mass shell that is gravitationally bound (i.e., with a negative total Newtonian energy) reaches a radius of maximum expansion (turnaround) and subsequently collapses. The solution of the equation of motion for a top-hat region shows that at the moment when the region collapses to a point, the overdensity predicted by linear theory is $\delta_L = 1.686$ in the $\Omega_m = 1$ case, with only a weak dependence on Ω_Λ in the more general case. Thus, a top hat would have collapsed at redshift z if its linear overdensity extrapolated to the present day (also termed the critical density of collapse) is

$$\delta_{\text{crit}}(z) = \frac{1.686}{D(z)}, \qquad (3.9)$$

where we set $D(z=0) = 1$.

Even a slight violation of the exact symmetry of the initial perturbation can prevent the top hat from collapsing to a point. Instead, the halo reaches a state of virial equilibrium through violent dynamical relaxation. We are familiar with the fact that the circular orbit of the Earth around the Sun has a kinetic energy which is half the magnitude of the gravitational potential energy. According to the *virial theorem*, this happens to be a property shared by all dynamically relaxed, self-gravitating systems. We may therefore use $U = -2K$ to relate the potential energy U to the kinetic energy K in the final state of a collapsed halo. This implies that the virial radius is half the turnaround radius (where the kinetic energy vanishes). Using this result, the final mean overdensity relative to ρ_c at the collapse redshift turns out to be $\Delta_c = 18\pi^2 \simeq 178$ in the $\Omega_m = 1$ case,* which applies at redshifts $z \gg 1$. We restrict our attention below to these high redshifts.

---

*This implies that dynamical time within the virial radius of galaxies, $\sim (G\rho_{\text{vir}})^{-1/2}$, is of order a tenth of the age of the Universe at any redshift.

A halo of mass M collapsing at redshift $z \gg 1$ thus has a virial radius

$$r_{\text{vir}} = 1.5 \left(\frac{M}{10^8 M_\odot} \right)^{1/3} \left(\frac{1+z}{10} \right)^{-1} \text{kpc}, \quad (3.10)$$

and a corresponding circular velocity

$$V_c = \left(\frac{GM}{r_{\text{vir}}} \right)^{1/2}$$

$$= 17.0 \left(\frac{M}{10^8 M_\odot} \right)^{1/3} \left(\frac{1+z}{10} \right)^{1/2} \text{km s}^{-1}.$$

$$(3.11)$$

We may also define a virial temperature

$$T_{\text{vir}} = \frac{\mu m_p V_c^2}{2k} = 1.04 \times 10^4 \left(\frac{\mu}{0.6} \right)$$

$$\times \left(\frac{M}{10^8 M_\odot} \right)^{2/3} \left(\frac{1+z}{10} \right) \text{K}, \quad (3.12)$$

where μ is the mean molecular weight and m_p is the proton mass. Note that the value of μ depends on the ionization fraction of the gas; for a fully ionized primordial gas $\mu = 0.59$, while a gas with ionized hydrogen but only singly ionized helium has $\mu = 0.61$. The binding energy of the halo is approximately

$$E_b = \frac{1}{2}\frac{GM^2}{r_{\text{vir}}}$$
$$= 2.9 \times 10^{53}\left(\frac{M}{10^8 M_\odot}\right)^{5/3}\left(\frac{1+z}{10}\right) \text{ ergs.}$$
$$\text{(3.13)}$$

Note that if the ordinary matter traces the dark matter, its total binding energy is smaller than E_b by a factor of Ω_b/Ω_m, and could be lower than the energy output of a single supernova* ($\sim 10^{51}$ ergs) for the first generation of dwarf galaxies.

Although spherical collapse captures some of the physics governing the formation of halos, structure formation in cold dark matter models proceeds hierarchically. At early times, most of the dark matter was in low-mass halos, and these halos then continuously accreted and merged to form high-mass halos. Numerical simulations of hierarchical halo formation indicate a roughly universal spherically averaged density profile for the resulting halos, though with considerable scatter among different halos. This profile has the form[†]

$$\rho(r) = \frac{3H_0^2}{8\pi G}(1+z)^3\Omega_m\frac{\delta_c}{c_N x(1+c_N x)^2}, \quad \text{(3.14)}$$

---

*A supernova is the explosion that follows the death of a massive star.

[†]This functional form is commonly labeled as the "NFW profile" after the original paper by J. F. Navarro, C. S. Frenk, and S.D.M. White, *Astrophys. J.* **490**, 493 (1997).

where $x = r/r_{\mathrm{vir}}$, and the characteristic density δ_c is related to the concentration parameter c_N by

$$\delta_c = \frac{\Delta_c}{3} \frac{c_N^3}{\ln(1 + c_N) - c_N/(1 + c_N)}. \qquad (3.15)$$

The concentration parameter itself depends on the halo mass M, at a given redshift z, with a value of order ~ 4 for newly collapsed halos.

3.4 Abundance of Dark Matter Halos

In addition to characterization of the properties of individual halos, a critical prediction of any theory of structure formation is the abundance of halos, namely, the number density of halos as a function of mass, at any redshift. This prediction is an important step toward inferring the abundances of galaxies and galaxy clusters. While the number density of halos can be measured for particular cosmologies in numerical simulations, an analytic model helps us gain physical understanding and can be used to explore the dependence of abundances on all the cosmological parameters.

A simple analytic model which successfully matches most of the numerical simulations was developed by Bill Press and Paul Schechter in 1974.[10] The model is based on the ideas of a Gaussian random field of density perturbations, linear gravitational growth, and spherical collapse. Once a region on the mass scale of interest reaches the threshold amplitude for a collapse according to linear theory, it can be declared as a virialized object. Counting the number of such density peaks per unit volume is straightforward for a Gaussian probability distribution.

To determine the abundance of halos at a redshift z, we use δ_M, the density field smoothed on a mass scale M, as defined in section 3.1. Since δ_M is distributed as a Gaussian variable with zero mean and standard deviation $\sigma(M)$ [which depends only on the present linear power spectrum; see equation (3.7)], the probability that δ_M is greater than some δ equals

$$\int_{\delta}^{\infty} d\delta_M \frac{1}{\sqrt{2\pi}\,\sigma(M)} \exp\left(-\frac{\delta_M^2}{2\,\sigma^2(M)}\right)$$
$$= \frac{1}{2}\text{erfc}\left(\frac{\delta}{\sqrt{2}\,\sigma(M)}\right). \qquad (3.16)$$

The basic ansatz is to identify this probability with the fraction of dark matter particles which are part of collapsed halos of mass greater than M at redshift z. There are two additional ingredients. First, the value used for δ is $\delta_{\text{crit}}(z)$ [given in equation (3.9)], which is the critical density of collapse found for a spherical top hat [extrapolated to the present since $\sigma(M)$ is calculated using the present power spectrum at $z=0$]; and second, the fraction of dark matter in halos above M is multiplied by an additional factor of 2 in order to ensure that every particle ends up as part of some halo with $M > 0$. Thus, the final formula for the mass fraction in halos above M at redshift z is

$$F(>M|z) = \text{erfc}\left(\frac{\delta_{\text{crit}}(z)}{\sqrt{2}\,\sigma(M)}\right). \qquad (3.17)$$

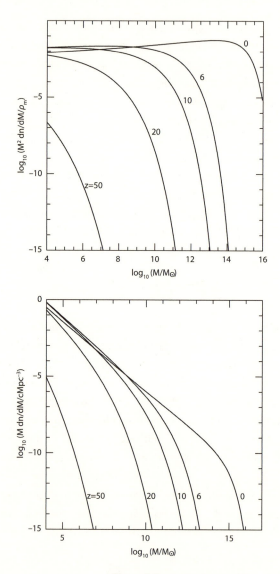

Figure 3.2.

Differentiating the fraction of dark matter in halos above M yields the mass distribution. Letting dn be the comoving number density of halos of mass between M and $M + dM$, we have

$$\frac{dn}{dM} = \sqrt{\frac{2}{\pi}} \frac{\rho_m}{M} \frac{-d(\ln \sigma)}{dM} \nu_c \, e^{-\nu_c^2/2}, \qquad (3.18)$$

where $\nu_c = \delta_{\mathrm{crit}}(z)/\sigma(M)$ is the number of standard deviations represented by the critical collapse overdensity on mass scale M. Thus, the abundance of halos depends on the two functions $\sigma(M)$ and $\delta_{\mathrm{crit}}(z)$, each of which depends on cosmological parameters. The above simple ansatz was refined over the years to provide a better match to numerical simulation. Results for the comoving density of halos of different masses at different redshifts are shown in figure 3.2.

The ad hoc factor of 2 in the Press-Schechter derivation is necessary, since otherwise only positive fluctuations of δ_M would be included. Bond et al. (1991) found an alternate derivation of this correction factor, using a

Figure 3.2. *Facing page, top:* The mass fraction incorporated into halos per logarithmic bin of halo mass $(M^2 dn/dM)/\rho_m$, as a function of M at different redshifts z. Here $\rho_m = \Omega_m \rho_c$ is the present-day matter density, and $n(M)dM$ is the comoving density of halos with masses between M and $M + dM$. The halo mass distribution was calculated based on an improved version of the Press-Schechter formalism for ellipsoidal collapse [R. K. Sheth and G. Tormen, *Mon. Not. R. Astron. Soc.* **329**, 61 (2002)] that better fits numerical simulations. *Bottom:* Number density of halos per logarithmic bin of halo mass, Mdn/dM (in units of comoving Mpc^{-3}), at various redshifts.

different ansatz, called the excursion set (or extended Press-Schechter) formalism.[11] In their derivation, the factor of 2 has a more satisfactory origin. For a given mass M, even if δ_M is smaller than $\delta_{crit}(z)$, it is possible that the corresponding region lies inside a region of some larger mass $M_L > M$, with $\delta_{M_L} > \delta_{crit}(z)$. In this case the original region should be counted as belonging to a halo of mass M_L. Thus, the fraction of particles which are part of collapsed halos of mass greater than M is larger than the expression given in equation (3.16).

The Press-Schechter formalism makes no attempt to deal with the correlations among halos or between different mass scales. This means that, while it can generate a distribution of halos at two different epochs, it says nothing about how particular halos in one epoch are related to those in the second. We therefore would like some method to predict, at least statistically, the growth of individual halos via accretion and mergers. Even if we restrict ourselves to spherical collapse, such a model must utilize the full spherically averaged density profile around a particular point. The potential correlations between the mean overdensities at different radii make the statistical description substantially more difficult.

The excursion set formalism seeks to describe the statistics of halos by considering the statistical properties of $\overline{\delta}(R)$, the average overdensity within some spherical window of characteristic radius R, as a function of R. While the Press-Schechter model depends only on the Gaussian distribution of $\overline{\delta}$ for one particular R, the excursion set considers all R. Again the connection between a value of the linear regime δ and the final state is made via the spherical collapse solution so that there is a critical value $\delta_{\mathrm{crit}}(z)$ of $\bar{\delta}$ which is required for collapse at a redshift z.

For most choices of window function, the functions $\bar{\delta}(R)$ are correlated from one R to another such that it is prohibitively difficult to calculate the desired statistics directly. However, for one particular choice of a window function, the correlations between different R greatly simplify, and many interesting quantities may be calculated.[12] The key is to use a k-space top-hat window function, namely, $W_k = 1$ for all k less than some critical k_c and $W_k = 0$ for $k > k_c$. This filter has a spatial form of $W(r) \propto j_1(k_c r)/k_c r$, which implies a comoving volume $6\pi^2/k_c^3$ or mass $6\pi^2 \rho_m/k_c^3$. The characteristic radius of the filter is $\sim k_c^{-1}$, as expected. Note that in real space this window function exhibits a sinusoidal oscillation and is not sharply localized.

The great advantage of the sharp k-space filter is that the difference at a given point between $\bar{\delta}$ on one mass scale and that on another mass scale is statistically independent from the value on the larger mass scale. With a Gaussian random field, each δ_k is Gaussian-distributed independently from the others. For this filter,

$$\bar{\delta}(M) = \int_{k<k_c(M)} \frac{d^3k}{(2\pi)^3} \delta_k, \qquad (3.19)$$

meaning that the overdensity on a particular scale is simply the sum of the random variables δ_k interior to the chosen k_c. Consequently, the difference between the $\bar{\delta}(M)$ on two mass scales is just the sum of the δ_k in the spherical k-shell between the two k_c, which is independent of the sum of the

δ_k interior to the smaller k_c. Meanwhile, the distribution of $\overline{\delta}(M)$, given no prior information, is still a Gaussian of mean zero and variance,

$$\sigma^2(M) = \frac{1}{2\pi^2} \int_{k < k_c(M)} dk \, k^2 P(k). \qquad (3.20)$$

If we now consider $\overline{\delta}$ as a function of scale k_c, we see that we begin from $\overline{\delta} = 0$ at $k_c = 0$ ($M = \infty$) and then add independently random pieces as k_c increases. This generates a random walk, albeit one whose step size varies with k_c. We then assume that, at redshift z, a given function $\overline{\delta}(k_c)$ represents a collapsed mass M corresponding to the k_c where the function first crosses the critical value $\delta_{\mathrm{crit}}(z)$. With this assumption, we may use the properties of random walks to calculate the evolution of the mass as a function of redshift.

It is now easy to rederive the Press-Schechter mass function, including the previously unexplained factor of 2. The fraction of mass elements included in halos of mass smaller than M is just the probability that a random walk remains below $\delta_{\mathrm{crit}}(z)$ for all k_c smaller than K_c, the filter cutoff appropriate to M. This probability must be the complement of the sum of the probabilies that (a) $\overline{\delta}(K_c) > \delta_{\mathrm{crit}}(z)$, or that (b) $\overline{\delta}(K_c) < \delta_{\mathrm{crit}}(z)$ but $\overline{\delta}(k'_c) > \delta_{\mathrm{crit}}(z)$ for some $k'_c < K_c$. But these two cases in fact have equal probability; any random walk belonging to class (a) may be reflected around its first upcrossing of $\delta_{\mathrm{crit}}(z)$ to produce a walk of class (b), and vice versa. Since the distribution of $\overline{\delta}(K_c)$ is simply Gaussian with variance $\sigma^2(M)$, the fraction of random walks falling into class (a) is simply

$(1/\sqrt{2\pi\sigma^2}) \int_{\delta_{\text{crit}}(z)}^{\infty} d\delta \, \exp\{-\delta^2/2\sigma^2(M)\}$. Hence, the fraction of mass elements included in halos of mass smaller than M at redshift z is simply

$$F(< M) = 1 - 2 \times \frac{1}{\sqrt{2\pi\sigma^2}} \int_{\delta_{\text{crit}}(z)}^{\infty} d\delta \, \exp\left\{-\frac{\delta^2}{2\sigma^2(M)}\right\},$$
$$(3.21)$$

which may be differentiated to yield the Press-Schechter mass function. We may now go further and consider how halos at one redshift are related to those at another redshift. If it is given that a halo of mass M_2 exists at redshift z_2, then we know that the random function $\bar{\delta}(k_c)$ for each mass element within the halo first crosses $\delta(z_2)$ at k_{c2} corresponding to M_2. Given this constraint, we may study the distribution of k_c where the function $\bar{\delta}(k_c)$ crosses other thresholds. It is particularly easy to construct the probability distribution for when trajectories first cross some $\delta_{\text{crit}}(z_1) > \delta_{\text{crit}}(z_2)$ (implying $z_1 > z_2$); clearly this occurs at some $k_{c1} > k_{c2}$. This problem reduces to the previous one if we translate the origin of the random walks from $(k_c, \bar{\delta}) = (0, 0)$ to $(k_{c2}, \delta_{\text{crit}}(z_2))$. We therefore find the distribution of halo masses M_1 that a mass element finds itself in at redshift z_1, given that it is part of a larger halo of mass M_2 at a later redshift z_2, is

$$\frac{dP}{dM_1}(M_1, z_1 | M_2, z_2)$$

$$= \sqrt{\frac{2}{\pi}} \frac{\delta_{\text{crit}}(z_1) - \delta_{\text{crit}}(z_2)}{[\sigma^2(M_1) - \sigma^2(M_2)]^{3/2}} \left|\frac{d\sigma(M_1)}{dM_1}\right|$$

$$\times \exp\left\{-\frac{[\delta_{\text{crit}}(z_1) - \delta_{\text{crit}}(z_2)]^2}{2[\sigma^2(M_1) - \sigma^2(M_2)]}\right\}. \quad (3.22)$$

This may be rewritten as saying that the quantity

$$\tilde{v} = \frac{\delta_{\mathrm{crit}}(z_1) - \delta_{\mathrm{crit}}(z_2)}{\sqrt{\sigma^2(M_1) - \sigma^2(M_2)}} \qquad (3.23)$$

is distributed as the positive half of a Gaussian with unit variance; equation (3.23) may be inverted to find $M_1(\tilde{v})$.

We can interpret the statistics of these random walks as those of merging and accreting halos. For a single halo, we may imagine that, as we look back in time, the object breaks into ever smaller pieces, similar to the branching of a tree. Equation (3.22) is the distribution of the sizes of these branches at some given earlier time. However, using this description of the ensemble distribution to generate random realizations of single merger trees has proven to be difficult. In all cases, one recursively steps back in time, at each step breaking the final object into two or more pieces. A simplified scheme may assume that at each time step the object breaks into only two pieces. One value from the distribution (3.22) then determines the mass ratio of the two branches.

We may also use the distribution of the ensemble to derive some additional analytic results. A useful example is the distribution of the epoch at which an object that has mass M_2 at redshift z_2 has accumulated half of its mass. The probability that the formation time is earlier than z_1 can be defined as the probability that at redshift z_1 a progenitor whose mass exceeds $M_2/2$ exists:

$$P(z_f > z_1) = \int_{M_2/2}^{M_2} \frac{M_2}{M} \frac{dP}{dM}(M, z_1 | M_2, z_2) dM,$$

$$(3.24)$$

where dP/dM is given in equation (3.22). The factor of M_2/M corrects the counting from being mass weighted to number weighted; each halo of mass M_2 can have only one progenitor of mass greater than $M_2/2$. Differentiation of equation (3.24) with respect to time gives the distribution of formation times. Overall, the excursion set formalism provides a good approximation to more exact numerical simulations of halo assembly and merging histories.

3.5 Cooling and Chemistry

When a dark matter halo collapses, the associated gas falls in at a speed of order V_c in equation (3.11). When multiple gas streams collide and settle to a static configuration, the gas shocks to the virial temperature T_{vir} in equation (3.12), at which it is held against gravity by its thermal pressure. In order for fragmentation to occur and stars to form, the collapsed gas has to cool and get denser until its Jeans mass drops to the mass scale of individual stars.

Cooling of the gas in the Milky Way galaxy (the so-called interstellar medium) is controlled by abundant heavy elements, such as carbon, oxygen, or nitrogen, which were produced in the interiors of stars. However, before the first stars formed there were no such heavy elements around and the gas was able to cool only through radiative transitions of atomic and molecular hydrogen. Figure 3.3 illustrates the cooling rate of the primordial gas as a function of its temperature. Below a temperature of $\sim 10^4$ K, atomic transitions are not effective because collisions among the atoms do not carry sufficient energy to

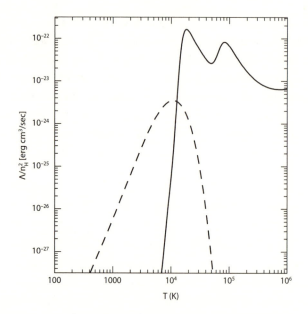

Figure 3.3. Cooling rates as a function of temperature for a primordial gas composed of atomic hydrogen and helium, as well as molecular hydrogen, in the absence of any external radiation. We assume a hydrogen number density $n_H = 0.045\,\mathrm{cm}^{-3}$, corresponding to the mean density of virialized halos at $z = 10$. The plotted quantity Λ/n_H^2 is roughly independent of density (unless $n_H > 10\,\mathrm{cm}^{-3}$), where Λ is the volume cooling rate (in erg/s/cm^3). The solid line shows the cooling curve for an atomic gas, with the characteristic peaks due to collisional excitation of hydrogen and helium. The dashed line shows the additional contribution of molecular cooling, assuming a molecular abundance equal to 1% of n_H.

excite the atoms and cause them to emit radiation through the decay of the excited states. Since the first gas clouds around the Jeans mass had a virial temperature well below 10^4 K, cooling and fragmentation of the gas had to rely on an alternative coolant with sufficiently low energy levels and a correspondingly low excitation temperature, namely, molecular hydrogen H_2. Hydrogen molecules could have formed through a rare chemical reaction involving the negative hydrogen (H^-) ion: $H + e^- \rightarrow H^- + photon$, and $H^- + H \rightarrow H_2 + e^-$, in which free electrons (e^-) act as catalysts. After cosmological recombination, the H_2 abundance was negligible. However, inside the first gas clouds, there was a sufficient abundance of free electrons to catalyze H_2 and cool the gas to temperatures as low as hundreds of Kelvins (similar to the temperature range presently on Earth).

However, the hydrogen molecule is fragile and can easily be broken by UV photons (with energies in the range of 11.26–13.6 eV),* to which the cosmic gas is transparent even before it is ionized.[13] The first population of stars was therefore suicidal. As soon as the very early stars formed and produced a background of UV light, this background light dissociated molecular hydrogen and suppressed the prospects for the formation of similar stars inside distant halos with low virial temperatures T_{vir}.

As soon as halos with $T_{vir} > 10^4$ K formed, atomic hydrogen was able to cool the gas in them and allow fragmentation even in the absence of H_2. In addition, once the gas was enriched with heavy elements, it was able to cool even more efficiently.

3.6 Sheets, Filaments, and Only Then, Galaxies

The development of large-scale cosmic structures occurs in three stages, as originally recognized by the Soviet physicist

---

* 1 electron-volt (eV) is an energy unit equivalent to 1.6×10^{-12} ergs or 11,604 K.

Figure 3.4. The large-scale distributions of dark matter (left) and gas (right) in the IGM show a network of filaments and sheets, known as the "cosmic web." Overall, the gas follows the dark matter on large scales but is more smoothly distributed on small scales owing to its pressure. The snapshots show the projected density contrast in a 7 Mpc thick slice at zero redshift from a numerical simulation of a box measuring 140 comoving Mpc on a side. Figure credit: H. Trac and U.-L. Pen, *New Astron.* **9**, 443 (2004).

Yakov Zel'dovich. First, a region collapses along one axis, making a two-dimensional sheet. Then the sheet collapses along the second axis, making a one-dimensional filament. Finally, the filament collapses along the third axis into a virialized halo. A snapshot of the distribution of dark matter at a given cosmic time should show a mix of these geometries in different regions that reached different evolutionary stages (owing to their different densities). The sheets define the boundary of voids from where their material was assembled; the intersection of sheets define filaments, and the intersection of filaments define halos— into which the material is ultimately drained. The resulting network of structures, shown in figure 3.4, delineates the so-called "cosmic web." Gas tends to follow the dark matter except within shallow potential wells into which it does not assemble, owing to its finite pressure. Computer simulations have provided highly accurate maps of how the dark matter is expected to be distributed since its dynamics is dictated only by gravity, but unfortunately, this matter is invisible. As soon as ordinary matter is added, complexity arises because of its cooling, chemistry, and fragmentation into stars and black holes. Although theorists have a difficult time modeling the dynamics of visible matter reliably, observers can monitor its distribution through telescopes. The art of cosmological studies of galaxies involves a delicate dance between what we observe but do not fully understand and what we fully understand but cannot observe.

Stars form in the densest, coolest knots of gas, in which the Jeans mass is lowered to the scale of a single star. By observing the radiation from galaxies, one is mapping the distribution of the densest peaks. The situation is analogous to a satellite image of the Earth at night in which light paints the special locations of big cities, while many other topographical details are hidden from view. It is, in principle, possible to probe the diffuse cosmic gas directly by observing its emission or absorption properties.

In the next three chapters we will describe two methods for studying structures in the early Universe: (i) imaging the stars and black holes within the first galaxies, and (ii) imaging the diffuse gas in between these galaxies.

4

THE FIRST STARS AND BLACK HOLES

There are two branches of theoretical research in cosmology. One considers the global properties of the Universe and the physical principles that govern it. As more data come in, our knowledge of the initial conditions as well as the underlying cosmological parameters gets refined with higher and higher precision. The second branch focuses on the formation of observable (luminous) objects out of the cosmic gas, including the stars and black holes in galaxies. Here, as more data come in, the models get more complex and the modelers understand more clearly why their previous analysis oversimplified the underlying processes. Theorists who work in the first branch run the risk of needing to switch fields in the future once the precision of the data becomes so good that there will be no point in further refinements (as happened in particle physics after its standard model was established in the 1970s). Theorists in the second branch run the risk of spending their career on a problem that will never get elegantly resolved.

The formation of the first stars hundreds of millions of years after the Big Bang marks the temporal boundary between these two branches. Earlier than that, the Universe was elegantly described by a small number of parameters. But as soon as the first stars formed, complex chemical and radiative processes entered the scene. 13.7 billion years later, we find very complex structures around us. Even though the present conditions in galaxies are a direct consequence of the simple initial conditions, the relationship between them was irreversibly blurred by complex processes over many decades of scales that cannot be fully simulated with present-day computers. Complexity reached its peak with the emergence of biology out of astrophysics. Although the journey that led to our existence was long and complicated, one fact is clear: our origins are traced to the production of the first heavy elements in the interiors of the first stars.

4.1 Metal-Free Stars

As mentioned in chapter 3, gas cooling in nearby galaxies is affected mostly by heavy elements (in a variety of forms, including atoms, ions, molecules, and dust) which are produced in stellar interiors and get mixed into the interstellar gas by supernova explosions. These powerful explosions are triggered at the end of the life of massive stars after their core consumes its nuclear fuel reservoir, loses its pressure support against gravity, and eventually collapses to make a black hole or a compact star made of neutrons with the density of an atomic nucleus. A neutron star has a size of order ~ 10 kilometers—comparable to a big city—but contains a mass comparable to the Sun. As infalling material arrives at the surface of the proto-neutron star, it bounces back and sends a shock wave into the surrounding envelope of the star which then explodes, exporting heavy elements into the surrounding medium.

The primordial gas out of which the first stars were made had 76% of its mass in hydrogen and 24% in helium, and did not contain elements heavier than lithium.[14] This is because, during Big-Bang nucleosynthesis, the cosmic expansion rate was too fast to allow the synthesis of heavier elements through nuclear fusion reactions. As a result, cooling of the primordial gas and its fragmentation into the first stars was initially mediated by trace amounts of molecular hydrogen in halos just above the cosmological Jeans mass of ~ 0.1–1 million solar masses ($T_{vir} \sim$ hundreds of Kelvins). Subsequently, star formation became much more efficient through the cooling of atomic hydrogen (see figure 3.3) in the first dwarf galaxies that were at least a thousand times more massive [$T_{vir} > 10^4$ K; see equation (3.12)]. The evolution of star formation in the first galaxies was also shaped by a variety of feedback processes. Internal self-regulation involved feedback from vigorous episodes of star formation (through supernova-driven winds) and black hole accretion (through the intense radiation and outflows it generates). But there was also external feedback. The reionization of the intergalactic gas heated the gas and elevated its Jeans mass. After reionization the intergalactic gas could not have assembled into the shallowest potential wells of dwarf galaxies.[15] This suppression of gas accretion may explain the inferred deficiency of dwarf galaxies relative to the much larger population of dark matter halos that is predicted to exist by numerical simulations but not observed around the Milky Way.[16] If this interpretation of the deficiency is correct, then most of the low-mass halos that formed after reionization were left devoid of gas and stars and are therefore invisible today. But before we get to these late stages, let us start at the beginning and examine the formation sites of the very first stars.

How did the the first clouds of gas form and fragment into the first stars? This questions poses a physics problem with well-specified initial conditions that can be solved on a computer. Starting with a simulation box in which primordial density fluctuations are realized (based on the initial power spectrum of density perturbations), one can simulate the collapse and fragmentation of the first gas clouds and the formation of stars within them.

Results from such numerical simulations of a collapsing halo with $\sim 10^6 M_\odot$ are presented in figure 4.1. Generically, the collapsing region makes a central massive clump with a typical mass of hundreds of solar masses, which happens to be the Jeans mass for a temperature of ~ 500 K and the density $\sim 10^4$ cm^{-3} at which the gas lingers because its H_2 cooling time is longer than its collapse time at that point. Soon after its formation, the clump becomes gravitationally unstable and undergoes runaway collapse at a roughly constant temperature due to H_2 cooling. The central clump does not typically undergo further subfragmentation and is expected to form a single star. Whether more than one star can form in a low-mass halo thus crucially depends on the degree of synchronization of clump formation,[17] since the radiation from the first star

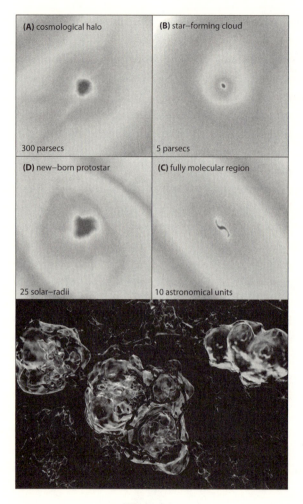

Figure 4.1.

to form can influence the motion of the surrounding gas more than gravity.[18]

How massive were the first stars? Star formation typically proceeds from the inside out, through the accretion of gas onto a central hydrostatic core. Whereas the initial mass of the hydrostatic core is very similar for primordial and present-day star formation, the accretion process—ultimately responsible for setting the final stellar mass—is expected to be rather different. On dimensional grounds, the mass growth rate is simply given by the ratio between the Jeans mass and the free-fall time, implying $(dM/dt) \sim c_s^3 / G \propto T^{3/2}$. A simple comparison of the temperatures in present-day star-forming regions, in which heavy elements cool the gas to a temperature as low as $T \sim 10$ K, with those

---

Figure 4.1. *Facing page:* Results from a numerical simulation of the formation of a metal-free star [N. Yoshida, K. Omukai, and L. Hernquist, *Science* **321**, 669 (2008)] and its feedback on its environment [V. Bromm, N. Yoshida, L. Hernquist, and C. F. McKee, *Nature* **459**, 49 (2009)]. *Top:* Projected gas distribution around a primordial protostar. Shown is the gas density (shaded so that dark gray denotes the highest density) of a single object on different spatial scales: (A) the large-scale gas distribution around the cosmological minihalo; (B) the self-gravitating, star-forming cloud; (C) the central part of the fully molecular core; and (D) the final protostar. *Bottom:* Radiative feedback around the first star involves ionized bubbles (light gray) and regions of high molecule abundance (medium gray). The large residual free electron fraction inside the relic ionized regions, left behind after the central star has died, rapidly catalyzes the reformation of molecules and a new generation of lower-mass stars.

in primordial clouds ($T \sim 200\text{--}300$ K) already indicates a difference in the accretion rate of more than two orders of magnitude. This suggests that the first stars were probably much more massive than their present-day analogs.

The rate of mass growth for the star typically tapers off with time.[19] A rough upper limit for the final mass of the star is obtained by continuing its accretion for its total lifetime of a few million years, yielding a final mass of $<10^3 \, M_\odot$. *Can a Population III star ever reach this asymptotic mass limit?* The answer to this question is not yet known with any certainty, and it depends on how accretion is eventually curtailed by feedback from the star.

The youngest stars in the Milky Way galaxy, with the highest abundance of elements heavier than helium (referred to by astronomers as "metals"), like the Sun, were historically categorized as Population I stars. Older stars, with much lower metallicity, were called Population II stars, and the first metal-free stars are referred to as Population III.

Currently, we have no direct observational constraints on how the first stars formed at the end of the cosmic dark ages, in contrast to the wealth of observational data we have on star formation in the local Universe.[20] Population I stars form out of cold, dense molecular gas that is structured in a complex, highly inhomogeneous way. The molecular clouds are supported against gravity by turbulent velocity fields and are pervaded on large scales by magnetic fields. Stars tend to form in clusters, ranging from a few hundred up to $\sim 10^6$ stars. It appears likely that the clustered nature of star formation leads to complicated dynamical interactions among the stars. The initial mass function

(IMF) of Population I stars is observed to have a broken power-law form, originally identified by Ed Salpeter,[21] with a number of stars $N_\star$ per logarithmic bin of star mass $M_\star$,

$$\frac{dN_\star}{d\log M_\star} \propto M_\star^{\beta}, \qquad (4.1)$$

where

$$\beta \simeq \begin{cases} -1.35 & \text{for } M_\star > 0.5 M_\odot, \\ 0.0 & \text{for } 0.008 M_\odot < M_\star < 0.5 M_\odot. \end{cases} \qquad (4.2)$$

The lower cutoff in mass corresponds roughly to the minimum fragment mass, set when the rate at which gravitational energy is released during the collapse exceeds the rate at which the gas can cool.[22] Moreover, nuclear fusion reactions do not ignite in the cores of protostars below a mass of $\sim 0.08 M_\odot$, so-called brown dwarfs. The most important feature of this IMF is that $\sim 1 M_\odot$ characterizes the mass scale of Population I star formation, in the sense that most of the stellar mass goes into stars with masses close to this value.

Since current simulations indicate that the first stars were predominantly very massive ($> 30 M_\odot$), and consequently rather different from present-day stellar populations, an interesting question arises: *how and when did the transition take place from the early formation of massive stars to the late-time formation of low-mass stars?*

The very first stars formed under conditions that were much simpler than the highly complex birth places of stars in present-day molecular clouds. As soon as these stars appeared, however, the situation became more complex due to their feedback on the environment. In particular, supernova explosions dispersed the heavy elements produced inside the first generation of stars into the surrounding gas. Atomic and molecular cooling became much more efficient after the addition of these metals.

Early metal enrichment was likely the dominant effect that brought about the transition from Population III to Population II star formation. Simple considerations indicate that as soon as the heavy element abundance exceeded a level as small as 0.1% (or even lower) of the solar abundance, the cooling of the gas became much more efficient than that provided by H_2 molecules.[23] The characteristic mass scale for star formation is therefore expected to be a function of metallicity, with a sharp transition at this metallicity threshold, above which the characteristic mass of a star gets reduced by about two orders of magnitude. Nevertheless, one should keep in mind that the temperature floor of the gas was dictated by the CMB (whose temperature was $54.6 \times [(1 + z)/20]$ K) and therefore, even with efficient cooling, the stars at high redshifts were likely more massive than the stars found today.

The maximum distance out to which a galactic outflow mixes heavy elements with the IGM can be estimated based on energy considerations. The mechanical energy released, E, will accelerate all the gas it encounters into a thin shell at a physical distance R_{max} from the central source. In doing so, it must accelerate the swept-up gas to the Hubble flow velocity at that distance, $v_s = H(z) R_{max}$. If the shocked gas has a short cooling time, then its original kinetic energy is lost and is unavailable for expanding the shell. Ignoring the gravitational effect of the host galaxy, deviations from the Hubble flow, and cooling inside the cavity bounded by the shell, energy conservation implies $E = \frac{1}{2} M_s v_s^2$, where $M_s = \frac{4\pi}{3} \bar{\rho} R_{max}^3$. At $z \gg 1$, this gives a maximum outflow distance $R_{max} \sim (50\,\text{kpc}) \times (E/10^{56}\,\text{ergs})^{1/5}[(1+z)/10]^{-6/5}$. The maximum radius of influence from a galactic outflow can therefore be estimated based on the total number of supernovae that power it (each releasing $\sim 10^{51}$ ergs, of which a substantial fraction may be lost by early cooling) or the mass M_{bh} of the central black hole (typically releasing a fraction of a percent of $M_{bh}c^2$ in mechanical energy). More detailed calculations give similar results.[24] Finally, the fraction of the IGM enriched with heavy elements can be obtained by multiplying the density of galactic halos with their individual volumes of influence.

Since the earliest galaxies represent high-density peaks and are therefore clustered, the metal enrichment process was inherently nonuniform. The early evolution of the volume filling of metals in the IGM can be inferred from the spectra of bright high-redshift sources.[25] Even at late cosmic times, it should be possible to find regions of the Universe that are composed of primordial gas and hence could make Population III stars. Since massive stars produce ionizing photons much more effectively than low-mass stars, the transition from Population III to Population II stars had important consequences for the ionization history of the cosmic gas. By a redshift of $z \sim 5$, the average metal abundance in the IGM is observed to be $\sim 1\%$ of the solar value, as expected from the heavy element yield of the same massive stars that reionized the Universe.[26]

4.2 Properties of the First Stars

Primordial stars that are hundreds of times more massive than the Sun have an effective surface temperature T_{eff} approaching $\sim 10^5$ K, with only a weak dependence on their mass. This temperature is ~ 17 times higher than the surface temperature of the Sun, $\sim 5,800$ K. These massive stars are held against their self-gravity by radiation pressure, having the so-called *Eddington luminosity* (see the derivation in section 4.3) which is strictly proportional to their mass $M_\star$,

$$L_E = 1.3 \times 10^{40} \left(\frac{M_\star}{100 M_\odot} \right) \text{ erg s}^{-1}, \qquad (4.3)$$

and is a few million times more luminous than the Sun, $L_\odot = 4 \times 10^{33} \text{ erg s}^{-1}$. Because of these characteristics, the total luminosity and color of a cluster of such stars simply depend on its total mass and not on the mass distribution of stars within it. The radii of these stars, $R_\star$, can be calculated by equating their luminosity to the emergent blackbody flux σT_{eff}^4 times their surface area $4\pi R_\star^2$ (where $\sigma = 5.67 \times 10^{-5} \text{ erg cm}^{-2} \text{ s}^{-1} \text{ deg}^{-4}$ is the Stefan-Boltzmann constant). This gives

$$R_\star = \left(\frac{L_E}{4\pi \sigma T_{\text{eff}}^4} \right)^{1/2} = 4.3 \times 10^{11} \text{ cm} \times \left(\frac{M_\star}{100 M_\odot} \right)^{1/2},$$
$$(4.4)$$

which is ~six times larger than the radius of the Sun, $R_\odot = 7 \times 10^{10}$ cm.

The high surface temperature of the first stars makes them ideal factories of ionizing photons. To free (ionize) an electron out of a hydrogen atom requires an energy of 13.6 eV (equivalent, through a division by Boltzmann's constant k_B, to a temperature of ~1.6×10^5 K), which is coincidentally the characteristic energy of a photon emitted by the first massive stars.

The first stars had lifetimes of a few million years, independent of their mass. During its lifetime, a Population III star produced ~10^5 ionizing photons per proton incorporated in it. This means that only a tiny fraction ($> 10^{-5}$) of all the hydrogen in the Universe needs to be assembled into Population III stars in order for there to be sufficient photons to ionize the rest of the cosmic gas. The actual required star formation efficiency depends on the fraction of all ionizing photons that escape from the host galaxies into the intergalactic space (f_{esc}) rather then being absorbed by hydrogen inside these galaxies.[27] For comparison, Population II stars produce on average ~ 4,000 ionizing photons per proton in them. If the Universe was ionized by such stars, then a much larger fraction (by a factor of ~25) of its gaseous content had to be converted into stars in order to have the same effect as Population III stars.

The first stars had a large impact on their gaseous environment. Their UV emission ionized and heated the surrounding gas, and their winds or supernova explosions pushed the gas around like a piston. Such feedback effects controlled the overall star formation efficiency within each galaxy, $f_\star$. This efficiency is likely to have been small in the earlier galaxies that had shallow potential wells, and it is possible that the subsequent Population II stars dominated the production of ionizing photons during reionization. By today, the global fraction of baryons converted into stars in the Universe is ~10%.[28]

The accounting of the photon budget required for reionization is simple. If only a fraction $f_\star \sim$ 10% of the gas in galaxies was converted into Population II stars and only $f_{esc} \sim$ 10% of the ionizing radiation escaped into intergalactic space, then more than ~$1/(4000 \times 10\% \times 10\%)$ = 2.5% of the matter in the Universe had to collapse into galaxies before there was one ionizing photon available per intergalactic hydrogen atom. Reionization completed once the number of ionizing photons grew by another factor of a few to compensate for recombinations in dense intergalactic regions.

It is also possible to predict the luminosity distribution of the first galaxies as a function of redshift and photon wavelength by "dressing up" the mass distribution of halos in figure 3.2 with light. The simplest prescription would be to assume that some fraction $f_\star(\Omega_b/\Omega_m)$ of the total mass in each halo above the Jeans mass is converted into stars with a prescribed stellar mass distribution (Population II or Population III) over some prescribed period of time (related to the rotation time of the disk). Using available computer codes for the combined spectrum of the stars as a function of time, one may then compute the luminosity distribution of the halos as a function of redshift and wavelength and make predictions for future observations.[29] The association of specific halo masses with galaxies of different luminosities can also be guided by their clustering properties.[30]

The end state in the evolution of massive Population III stars depends on their mass. Within an intermediate mass range of $(140–260) M_\odot$ they were likely to explode as energetic *pair-instability supernovae*, and outside this mass range they were likely to implode into black holes. A pair-instability supernova is triggered when the core of the very massive low-metallicity star heats up in the last stage of its evolution. This leads to the production of electron-positron pairs as a result of collisions between atomic nuclei and energetic gamma rays, which in turn reduces thermal pressure inside the star's core. The pressure drop leads to a partial collapse and then greatly accelerated burning in a runaway thermonuclear explosion which blows the star up without leaving a remnant behind. The kinetic energy released in the explosion could reach $\sim 10^{53}$ ergs, exceeding the kinetic energy output of typical supernovae by two orders of magnitude. Although the characteristics of these powerful explosions were predicted theoretically several decades ago, there has been no conclusive evidence for their existence so far. Because of their exceptional energy outputs, pair-instability supernovae would be prime targets for future surveys of the first stars with the next generation of telescopes (see chapter 6).

Where are the remnants of the first stars located today? The very first stars formed in rare high-density peaks, hence their black hole remnants are likely to populate the cores of present-day galaxies. However, the bulk of the stars which formed in low-mass systems at later times are expected to behave similarly to the collisionless dark matter particles, and populate galaxy halos.

Although the very first generation of local galaxies are buried deep in the core of the Milky Way, most of the stars there today formed much later, making the search for rare old stars as impractical as finding needles in a haystack. Because the outer Milky Way halo is far less crowded with younger stars, it is much easier to search for old stars there. Existing halo surveys discovered a population of stars with exceedingly low iron abundance ($\sim 10^{-5}$ of the solar abundance of iron relative to hydrogen), but these "anemic" stars have a high abundance of other heavy elements, such as carbon.[31] We do not expect to find the very first population of massive stars in these surveys, since these had a lifetime of only a few million years, several orders of magnitude shorter than the period of time that has elapsed since the dark ages.

4.3 The First Black Holes and Quasars

A black hole is the end product from the complete gravitational collapse of a material object, such as a massive star. It is surrounded by a horizon from which even light cannot escape. Black holes have the dual virtues of being extraordinarily simple solutions to Einstein's equations of gravity (as they are characterized only by their mass, charge, and spin), but also the most disparate from their Newtonian analogs. In Einstein's theory, black holes represent the ultimate prisons: you can check in, but you can never check out.

Ironically, black hole environments are the brightest objects in the universe. Of course, it is not the black hole that is shining, but rather the surrounding gas is heated by viscously rubbing against itself and shining as it spirals into the black hole like water going down a drain, never to be seen again. The origin of the radiated energy is the release of gravitational binding energy as the gas falls into the deep gravitational potential well of the black hole. As much as tens of percent of the mass of the accreting material can be converted into heat (more than an order of magnitude beyond the maximum efficiency of nuclear fusion). Astrophysical black holes appear in two flavors: stellar-mass black holes that form when massive stars die, and the monstrous supermassive black holes that sit at the center of galaxies, reaching masses of up to 10 billion Suns. The latter type are observed as quasars and active galactic nuclei (AGN). It is by studying these accreting black holes that all of our observational knowledge of black holes has been obtained.

If this material is organized into a thin accretion disk, where the gas can efficiently radiate its released binding energy, then its theoretical modeling is straightforward. Less well understood are radiatively inefficient accretion flows, in which the inflowing gas obtains a thick geometry. It is generally unclear how gas migrates from large radii to near the horizon and how, precisely, it falls into the black hole. We presently have very poor constraints on how magnetic fields embedded and created by the accretion flow are structured, and how that structure affects the observed properties of astrophysical black holes. While it is beginning to be possible to perform computer simulations

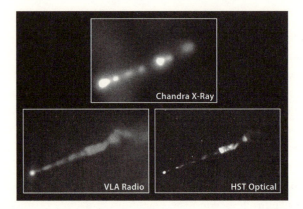

Figure 4.2. Multiwavelength images of the highly collimated jet emanating from the supermassive black hole at the center of the giant elliptical galaxy M87. The X-ray image (top) was obtained with the Chandra X-ray satellite, the radio image (bottom left) was obtained with the Very Large Array (VLA), and the optical image (bottom right) was obtained with the Hubble Space Telescope (HST).

of the entire accreting region, we are decades away from true *ab initio* calculations, and thus observational input plays a crucial role in deciding between existing models and motivating new ideas.

More embarrassing is our understanding of black hole jets (see figure 4.2). These extraordinary exhibitions of the power of black holes are moving at nearly the speed of light and involve narrowly collimated outflows whose base has a size comparable to the solar system, while their front reaches scales comparable to the distance between galaxies.[32] Unresolved issues are as basic as what jets are made of (whether electrons and protons or electrons and positrons, or primarily electromagnetic fields) and how they are accelerated in the first place. Both of these rest critically on the role of the black hole spin in the jet-launching process.

A quasar is a pointlike ("quasi-stellar") bright source at the center of a galaxy. There are many lines of evidence indicating that a quasar involves a supermassive black hole, weighting up to ten billion Suns, which is accreting gas from the core of its host galaxy. The supply of large quantities of fresh gas is often triggered by a merger between two galaxies. The infalling gas heats up as it spirals toward the black hole and dissipates its rotational energy through viscosity. The gas is expected to be drifting inward in an accretion disk whose inner "drain" has the radius of the *innermost stable circular orbit* (ISCO), according to Einstein's theory of gravity. Interior to the ISCO, the gas plunges into the black hole in such a short time that it has no opportunity to radiate most of its thermal energy. However, the fraction of the rest mass of the gas which gets radiated away just outside the ISCO is high, ranging between 5.7% for a nonspinning black hole and 42% for a maximally spinning black hole.[33] This "radiative efficiency" is far greater than the mass-energy conversion efficiency provided by nuclear fusion in stars, which is < 0.7%.

Quasar activity is observed in a small fraction of all galaxies at any cosmic epoch. Mammoth black holes weighing more than a billion solar masses were discovered at redshifts as high as $z \sim 6.5$, less than a billion years after the Big Bang. *If massive black holes grew at early cosmic times, should their remnants be around us today?* Indeed, searches for black holes in local galaxies have found that every galaxy with a stellar spheroid harbors a supermassive black hole at its center. This implies that quasars are rare simply because their activity is short lived. Moreover, there appears to be a tight correlation between the black hole mass and the gravitational potential-well depth of their host spheroids of stars (as measured by the velocity dispersion of these stars). This suggests that the black holes grow up to the point where the heat they deposit into their environment or the piston effect from their winds prevent additional gas from feeding them further. The situation is similar to a baby who gets more energetic as he eats more at the dinner table, until his hyperactivity is so intense that he pushes the food off the table and cannot eat any more. This *principle of self-regulation* explains why quasars are short lived and why the final black hole mass is dictated by the depth of the potential in which the gas feeding it resides.[34] Most black holes today are dormant or "starved" because the gas around them was mostly used up in making the stars, or because their activity heated or pushed it away a long time ago.

What seeded the formation of supermassive black holes only a billion years after the Big Bang? We know how to make a black hole out of a massive star. When the star ends its life, it stops producing sufficient energy to hold itself against its own gravity, and its core collapses to make a black hole. Long before evidence for black holes was observed, this process leading to their existence was understood theoretically by Robert Oppenheimer and Hartland Snyder in 1937. However, growing a supermassive black hole is more difficult. There is a maximum luminosity at which the environment of a black hole of mass M_{BH} may shine and still accrete gas.* This Eddington luminosity, L_E, is obtained from balancing the inward force of gravity on each proton by the outward radiation force on its companion electron at a distance r:

$$\frac{GM_{BH}m_p}{r^2} = \frac{L_E}{4\pi r^2 c}\sigma_T, \qquad (4.5)$$

where m_p is the proton mass and $\sigma_T = 0.67 \times 10^{-24}$ cm^2 is the cross section for scattering a photon by an electron. Interestingly, the limiting luminosity is independent of radius in the Newtonian regime. Since the Eddington luminosity represents an exact balance between gravity and radiation forces, it actually equals to the luminosity of massive stars which are held at rest against gravity by radiation pressure, as described by equation (4.3). This limit is formally valid in a spherical geometry, and exceptions to it were conjectured for other accretion geometries over the years. But, remarkably, observed quasars for which black hole masses can be measured by independent methods appear to respect this limit.

---

*Whereas the gravitational force acts mostly on the protons, the radiation force acts primarily on the electrons. These two species are tied together by a global electric field, so that the entire "plasma" (ionized gas) behaves as a single quasi-neutral fluid which is subject to both forces. Under similar circumstances, electrons are confined to the Sun by an electric potential of about a kilovolt (corresponding to a total charge of ∼75 coulombs). The opposite electric forces per unit volume acting on electrons and ions in the Sun cancel out so that the total pressure force is exactly balanced by gravity, as for a neutral fluid. An electric potential of 1–10 kilovolts also binds electrons to clusters of galaxies (where the thermal velocities of these electrons, ∼0.1c, are well in excess of the escape speed from the gravitational potential). For a general discussion, see A. Loeb, *Phys. Rev. D* **37**, 3484 (1988).

The total luminosity from gas accreting onto a black hole, L, can be written as some radiative efficiency ϵ times the mass accretion rate $\dot{M}$,

$$L = \epsilon \dot{M} c^2, \qquad (4.6)$$

with the black hole accreting the nonradiated component, $\dot{M}_{BH} = (1 - \epsilon)\dot{M}$. The equation that governs the growth of the black hole mass is then

$$\dot{M}_{BH} = \frac{M_{BH}}{t_E}, \qquad (4.7)$$

where (after substituting all fundamental constants),

$$t_E = (4 \times 10^7 \text{ yr}) \left(\frac{\epsilon/(1-\epsilon)}{10\%} \right) \left(\frac{L}{L_E} \right)^{-1}. \qquad (4.8)$$

We therefore find that, as long as fuel is amply supplied, the black hole mass grows exponentially in time, $M_{BH} \propto \exp\{t/t_E\}$, with an e-folding time t_E. Since the growth time in equation (4.8) is significantly shorter than the $\sim 10^9$ years corresponding to the age of the Universe at a redshift $z \sim 6$, where black holes with a mass $\sim 10^9 M_\odot$ are found, one might naively conclude that there is plenty of time to grow the observed black hole masses from small seeds. For example, a seed black hole from a Population III star of $100 M_\odot$ can grow in less than a billion years up to $\sim 10^9 M_\odot$ for $\epsilon \sim 10\%$ and $L \sim L_E$. However, the intervention of various processes makes it unlikely that a stellar mass seed will be able to accrete continuously at its Eddington limit with no interruption.

For example, mergers are very common in the early Universe. Every time two gas-rich galaxies come together, their black holes are likely to coalesce. The coalescence is initially triggered by "dynamical friction" on the surrounding gas and stars, and is completed—when the binary gets tight—as a result of the emission of gravitational radiation.[35] The existence of gravitational waves is a generic prediction of Einstein's theory of gravity. They represent ripples in space-time generated by the motion of the two black holes as they move around their common center of mass in a tight binary. The energy carried by the waves is taken away from the kinetic energy of the binary, which therefore gets tighter with time. Computer simulations reveal that, when two black holes with unequal masses merge to make a single black hole, the remnant gets a kick due to the nonisotropic emission of gravitational radiation at the final plunge.* This kick was calculated recently using advanced computer codes that solve Einstein's equations (a task that was plagued for decades with numerical instabilities).[36] The typical kick velocity is hundreds of kilometers per second (and up to ten times more for special spin orientations), bigger than the escape speed from the first dwarf galaxies.[37] This implies that continuous accretion was likely punctuated by black hole ejection events,[38] forcing the merged dwarf galaxy to grow a new black hole seed from scratch.†

---

*The gravitational waves from black hole mergers at high redshifts could in principle be detected by a proposed space-based mission called the *Laser Interferometer Space Antenna* (LISA). For more details, see http://lisa.nasa.gov/, and, for example, J.S.B. Wyithe and A. Loeb, *Astrophys. J.* **590**, 691 (2003).

†These black hole recoils might have left observable signatures in the local Universe. For example, the halo of the Milky Way galaxy may include hundreds of freely floating ejected black holes with compact star clusters around them,

If continuous feeding is halted, or if the black hole is temporarily removed from the center of its host galaxy, then one is driven to the conclusion that the black hole seeds must have started more massive than $\sim 100 M_\odot$. More massive seeds may originate from supermassive stars. *Is it possible to make such stars in early galaxies?* Yes, it is. Numerical simulations indicate that stars weighing up to a million Suns could have formed at the centers of early dwarf galaxies which were barely able to cool their gas through transitions of atomic hydrogen, having $T_{\text{vir}} \sim 10^4$ K and no H_2 molecules. Such systems have a total mass that is several orders of magnitude higher than the earliest Jeans-mass condensations discussed in section 4.1. In both cases, the gas lacks the ability to cool well below T_{vir}, and so it fragments into one or two major clumps. The simulation shown in figure 4.3 results in clumps of several million solar masses, which inevitably end up as massive black holes. The existence of such seeds would have given a jump start to the black hole growth process.

The nuclear black holes in galaxies are believed to be fed with gas in episodic events of gas accretion triggered by mergers of galaxies. The energy released by the accreting gas during these episodes could easily unbind the gas reservoir from the host galaxy and suppress star formation within it. If so, nuclear black holes regulate their own growth by expelling the gas that feeds them. In so doing, they also shape the stellar content of their host galaxy. This

representing relics of the early mergers that assembled the Milky Way out of its original building blocks of dwarf galaxies [R. O'Leary and A. Loeb, *Mon. Not. R. Astron. Soc.* **395**, 781 (2009)].

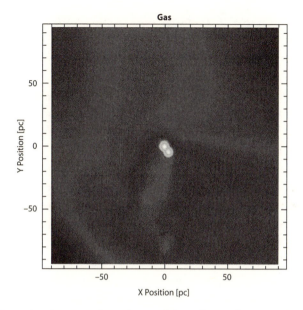

Figure 4.3. Numerical simulation of the collapse of an early dwarf galaxy with a virial temperature just above the cooling threshold of atomic hydrogen and no H_2. The image shows a snapshot of the gas density distribution 500 million years after the Big Bang, indicating the formation of two compact objects near the center of the galaxy with masses of $2.2 \times 10^6 M_\odot$ and $3.1 \times 10^6 M_\odot$, respectively, and radii <1 pc. Subfragmentation into lower-mass clumps is inhibited because hydrogen atoms cannot cool the gas significantly below its initial temperature. These circumstances lead to the formation of supermassive stars that inevitably collapse to make massive seeds of supermassive black holes. The simulated box size is 200 pc on a side. Figure credit: V. Bromm and A. Loeb, *Astrophys. J.* **596**, 34 (2003).

may explain the observed tight correlations between the mass of central black holes in present-day galaxies and the velocity dispersion[39] $\sigma_\star$ or luminosity[40] L_{sp} of their host spheroids of stars (namely, $M_{\mathrm{BH}} \propto \sigma_\star^4$ or $M_{\mathrm{BH}} \propto L_{\mathrm{sp}}$). Since the mass of a galaxy at a given redshift scales with its virial velocity as $M \propto V_c^3$ in equation (3.11), the binding energy of galactic gas is expected to scale as $M V_c^2 \propto V_c^5$ while the momentum required to kick the gas out of its host would scale as $M V_c \propto V_c^4$. Both scalings can be tuned to explain the observed correlations between black hole masses and the properties of their host galaxies.[41] Star formation inevitably precedes black hole fueling, since the outer region of the accretion flows that feed nuclear black holes is typically unstable to fragmentation.[42] This explains the high abundance of heavy elements inferred from the broad emission lines of quasars at all redshifts.[43]

The inflow of cold gas toward galaxy centers during the growth phase of their black holes would naturally be accompanied by a burst of star formation. The fraction of gas not consumed by stars or ejected by supernova-driven winds will continue to feed the black hole. It is therefore not surprising that quasar and star-burst activities coexist in ultraluminous galaxies, and that all quasars show strong spectral lines of heavy elements. Similarly to the above-mentioned prescription for modeling galaxies, it is possible to "dress up" the mass distribution of halos in figure 3.2 with quasar luminosities (related to L_E, which is a prescribed function of M based on the observed M_{BH}-$\sigma_\star$ relation) and a duty cycle (related to t_E), and find the evolution of the quasar population over redshift. This simple approach can be tuned to give good agreement with existing data on quasar evolution.[44]

The early growth of massive black holes led to the supermassive black holes observed today. In our own

Milky Way galaxy, stars are observed to zoom around the Galactic center at speeds of up to ten thousand kilometers per second, owing to the strong gravitational acceleration near the central black hole.[45] But closer-in observations are forthcoming. Existing technology should soon be able to image the silhouette of the supermassive black holes in the Milky Way and M87 galaxies directly (see figure 4.4).

4.4 Gamma-Ray Bursts: The Brightest Explosions

Gamma-ray bursts (GRBs) were discovered in the late 1960s by the American Vela satellites, built to search for flashes of high-energy photons ("gamma rays") from Soviet nuclear weapon tests in space. The United States suspected that the Soviets might attempt to conduct secret nuclear tests after signing the Nuclear Test Ban Treaty in 1963. On July 2, 1967, the Vela 4 and Vela 3 satellites detected a flash of gamma radiation unlike any known nuclear weapons signature. Uncertain of its meaning but not considering the matter particularly urgent, the team at the Los Alamos Laboratory, led by Ray Klebesadel, filed the data away for future investigation. As additional Vela satellites were launched with better instruments, the Los Alamos team continued to find unexplained GRBs in their data. By analyzing the different arrival times of the bursts as detected by different satellites, the team was able to estimate the sky positions of 16 bursts and definitively rule out either a terrestrial or solar origin. The discovery was declassified and published in 1973 (*Astrophys. J.* **182**,

Figure 4.4. Simulated image of an accretion flow around a black hole spinning at half its maximum rate, from a viewing angle of 10° relative to the rotation axis. The coordinate grid in the equatorial plane of the spiraling flow shows how strong lensing around the black hole bends the back of the apparent disk up. The left side of the image is brighter due its rotational motion toward the observer. The bright arcs are generated by gravitational lensing. A dark silhouette appears around the location of the black hole because the light emitted by gas behind it disappears into the horizon and cannot be seen by an observer on the other side. Recently, the technology for observing such an image from the supermassive black holes at the centers of the Milky Way and M87 galaxies has been demonstrated as feasible [S. Doeleman et al., *Nature* **455**, 78 (2008)]. To obtain the required resolution of tens of micro-arcseconds, signals are being correlated over an array (interferometer) of observatories operating at a millimeter wavelength across the Earth. Figure credit: A. Broderick and A. Loeb, *J. Phys. Conf. Ser.* **54**, 448 (2006); *Astrophys. J.* **697**, 1164 (2009).

L85) under the title "Observations of Gamma-Ray Bursts of Cosmic Origin."

The distance scale and nature of GRBs remained mysterious for more than two decades. Initially, astronomers favored a local origin for the bursts, associating them with sources within the Milky Way. In 1991, the Compton Gamma Ray Observatory satellite was launched, and its "Burst and Transient Source Explorer" instrument started to discover a GRB every day or two, increasing the total number of known GRBs up to a few thousand. The larger statistical sample of GRBs made it evident that their distribution on the sky is isotropic. Such a distribution would be most natural if the bursts originate at cosmological distances since the Universe is the only system which is truly isotropic around us. Nevertheless, the local origin remained more popular within the GRB community for six years, until February 1997, when the Italian-Dutch satellite BeppoSAX detected a gamma-ray burst (GRB 970228) and localized it to within minutes of arc using its X-ray camera. With this prompt localization, ground-based telescopes were able to identify a fading counterpart in the optical band. Once the GRB afterglow faded, deep imaging revealed a faint, distant host galaxy at the location of the optical afterglow of the GRB. The association of a host galaxy at a cosmological distance for this burst and many subsequent ones revised the popular view in favor of associating GRBs with cosmological distances. This shift in popular view provides testimony to how a psychological bias in the scientific community can be overturned by hard scientific evidence.[46]

A GRB afterglow is initially brightest at short photon wavelengths and then fades away at longer wavelengths, starting in the X-ray band (over time scales of minutes to hours), shifting to the UV and optical band (over days), and ending in the infrared and radio (over weeks and months).* Among the first detected afterglows, observers noticed that as the afterglow lightcurve faded, long-duration GRBs showed evidence for a supernova flare, indicating that they are also associated with core-collapse supernova events. The associated supernovae were classified as related to massive stars which have lost their hydrogen envelope in a wind. In addition, long-duration GRBs were found to be associated with star-forming regions where massive stars form and explode only a million years after being born. These clues indicated that long-duration GRBs are most likely associated with massive stars. The most popular model for long-duration GRBs became known as the "collapsar" model[47] (see figure 4.5). According to this model, the progenitor of the GRB is a massive star whose core eventually consumes its nuclear fuel, loses pressure support, and collapses. If the core of the star is too massive to make a neutron star, it collapses to a black hole. As material is spiraling into the black hole, two jets are produced at a speed close to that of light. So far, there is nothing spectacular about this setting, since we see scaled-up versions of such jets being formed around massive black holes in the centers of galaxies, as

---

*For an extreme example of a GRB afterglow from a redshift $z = 0.94$ that was bright enough to be seen with the naked eye, see J. Bloom et al., *Astrophys. J.* **691**, 723 (2009).

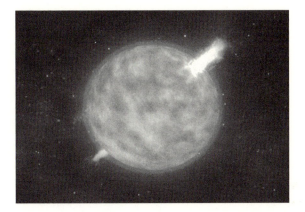

Figure 4.5. Illustration of a long-duration gamma-ray burst in the popular "collapsar" model. The collapse of the core of a massive star (which lost its hydrogen envelope) to a black hole generates two opposite jets moving out at a speed close to the speed of light. The jets drill a hole in the star and shine brightly toward an observer who happens to be located within the collimation cones of the jets. The jets emanating from a single massive star are so bright that they can be seen across the Universe out to the epoch when the first stars formed. Figure credit: NASA E/PO.

shown in figure 4.2. However, when jets are generated in the core of a star, they have to make their way out by drilling a hole in the surrounding dense envelope. As soon as the head of a jet exits, the highly collimated stream of radiation emanating from it would appear as a gamma-ray flash to an observer who happened to line up with its jet axis. The subsequent afterglow results from the interaction between the jet and the ambient gas in the vicinity of the progenitor star. As the jet slows down by pushing against the ambient medium, the nonthermal radiation from accelerated relativistic electrons in the shock wave in front of it gets shifted to longer wavelengths and fainter luminosities. Also, as the jet makes its way out of the star, its piston effect deposits energy in the stellar envelope and explodes the star, supplementing the GRB with a supernova-like explosion. Because of their immense luminosities, GRBs can be observed out to the edge of the Universe. These bright signals may be thought of as the cosmic fireworks signaling the birth of black holes at the end of the life of their parent massive stars. If the first stars produced GRBs (as their descendants do in the more recent Universe), then they would be detectable out to their highest redshifts. Their powerful beacons of light can be used to illuminate the dark ages and probe the cosmic gas around the time when it condensed to make the first galaxies. As this book was written, a gamma-ray burst was discovered by the Swift Satellite[48] at a redshift 8.3, representing the most distant source known, originating at the time when the Universe was only ~620 million years old.[49]

5

THE REIONIZATION OF COSMIC HYDROGEN BY THE FIRST GALAXIES

5.1 Ionization Scars by the First Stars

The cosmic microwave background (CMB) indicates that hydrogen atoms formed 400 thousand years after the Big Bang, as soon as the gas cooled below 3,000 K as a result of cosmological expansion. Observations of the spectra of early galaxies, quasars, and gamma-ray bursts indicate that less than a billion years later the same gas underwent a wrenching transition from atoms back to their constituent protons and electrons in a process known as reionization. Indeed, the bulk of the Universe's ordinary matter today is in the form of free electrons and protons, located deep in intergalactic space. The free electrons have other side effects; for example, they scatter the CMB and produce polarization fluctuations at large angles on the sky.* The latest analysis of the CMB polarization data from WMAP[50] indicates that $(8.7 \pm 1.7)\%$ of the CMB photons were scattered by free electrons after cosmological recombination, implying that reionization took place at around a redshift $z \sim 10$, only 500 million years after the Big Bang.[†] It is intriguing that the inferred reionization epoch coincides with the appearance of the first galaxies, which inevitably produced ionizing radiation. *How was the primordial gas transformed to an ionized state by the first galaxies within merely hundreds of million of years?*

We can address this question using the results of chapters 3 and 4 concerning the formation rate of new galaxies at various cosmic epochs. The course of reionization can be determined by counting photons from all galaxies as a function of time. Both stars and black holes contribute ionizing photons, but the early Universe is dominated by small galaxies which, in the local universe, have disproportionately small central black holes. In fact, bright quasars are known to be extremely rare above redshift 6,

---

*Polarization is produced when free electrons scatter a radiation field with a quadrupole anisotropy Q. Consequently, reionization generates a fractional CMB polarization of $P \sim 0.1\tau Q$ out of the CMB quadrupole on scales of the horizon at reionization. Here, $\tau = \int_0^{z_{\text{reion}}} n_e(z)\sigma_T(c\,dt/dz)dz$ is the optical depth for scattering by electrons of mean density $n_e(z)$, which provides a measure of the redshift of reionization, z_{reion}. (The scattering probability, $\tau \ll 1$, is the chance that a photon will encounter an electron within the volume associated with the scattering cross section σ_T times the photon path length $\int c\,dt$.) For a sudden reionization of hydrogen and neutral helium at redshift z_{reion}, $\tau = 4.75 \times 10^{-3} \times \{[\Omega_\Lambda + \Omega_m(1 + z_{\text{reion}})^3]^{1/2} - 1\}$.

[†] This number will be refined by forthcoming CMB data from the Planck satellite (http://www.rssd.esa.int/index.php?project=planck), and will be supplemented by constraints from 21 cm observations as described in chapter 7 [see J. Pritchard, A. Loeb, and J.S.B. Wyithe, http://arxiv.org/abs/0908.3891 (2009)].

indicating that stars most likely dominated the production of ionizing UV photons during the reionization epoch.[51] Since stellar ionizing photons are only slightly more energetic than the 13.6 eV ionization threshold of hydrogen, they are absorbed efficiently once they reach a region with substantial neutral hydrogen. This makes the intergalactic medium (IGM) during reionization a two-phase medium characterized by highly ionized regions separated from neutral regions by sharp ionization fronts. We can obtain a first estimate of the requirements of reionization by demanding one stellar ionizing photon for each hydrogen atom in the Universe. As discussed in section 4.2, if we conservatively assume that stars within the early galaxies were similar to those observed locally, then each star produced $\sim 4,000$ ionizing photons per proton in it. Star formation is observed today to be an inefficient process, but even if stars in galaxies formed out of only a fraction $f_\star \sim 10\%$ of the available gas, this amount was still sufficient to assemble only a small fraction of the total mass in the universe into galaxies in order to ionize the entire IGM. This fraction would be $\sim 2.5\%(f_{esc}/10\%)^{-1}$ for an escape fraction f_{esc} of ionizing UV photons out of galaxies. More detailed estimates of the actual required fraction account for the formation of some Population III stars (which were more efficient ionizers, as discussed in chapter 4), and for recombinations of hydrogen atoms at high redshifts and in dense regions. At the mean IGM density, the recombination time of hydrogen is shorter than the age of the Universe at $z > 8$.

From studies of the processed spectra of quasars at $z \sim 6$, we know that the IGM is highly ionized a billion years after the Big Bang. There are hints, however, that some large neutral hydrogen regions persist at these early times, which suggests that we may not need to go to much higher redshifts to begin to see the epoch of reionization. We now know that the universe could not have fully reionized earlier than an age of 300 million years, since WMAP observed the effect of the freshly created plasma at reionization on the large-scale polarization anisotropies of the CMB which limits the reionization redshift; an earlier reionization, when the universe was denser, would have created a stronger scattering signature that would have been inconsistent with the WMAP observations. In any case, the redshift at which reionization ended only constrains the overall cosmic efficiency for producing ionizing photons. In comparison, a detailed picture of reionization in progress will teach us a great deal about the population of the first galaxies that produced this cosmic phase transition.

5.2 Propagation of Ionization Fronts

Astronomers label the neutral and singly ionized states of an atomic species as I and II; for example, abbreviating neutral hydrogen as H I and ionized hydrogen as H II. The radiation output from the first stars ionizes H I in a growing volume, eventually encompassing almost the entire IGM within a single H II bubble. In the early stages of this process, each galaxy produced a distinct H II region, and only when the overall H II filling factor became significant did neighboring bubbles begin to overlap in large numbers, ushering in the "overlap phase" of reionization. Thus, the first goal of a model of reionization is to describe the initial stage, during which each source produces an isolated expanding H II region.

Let us consider, for simplicity, a spherical ionized volume V, separated from the surrounding neutral gas by a sharp ionization front. In the absence of recombinations, each hydrogen atom in the IGM would only have to be ionized once, and the ionized physical volume V_p would simply be determined by

$$\bar{n}_H V_p = N_\gamma, \qquad (5.1)$$

where $\bar{n}_H$ is the mean number density of hydrogen and N_γ is the total number of ionizing photons produced by the source. However, the elevated density of the IGM at high redshift implies that recombinations cannot be ignored. Just before World War II, the Danish astronomer Bengt Strömgren analyzed the same problem for hot stars embedded in the interstellar medium.[52] In the case of a steady ionizing source (and neglecting the cosmological expansion), he found that a steady-state volume (now termed a "Strömgren sphere") would be reached, through which recombinations are balancing ionizations:

$$\alpha_B \bar{n}_H^2 V_p = \frac{dN_\gamma}{dt}, \qquad (5.2)$$

where the recombination rate depends on the square of the density and on the recombination coefficient* (to all

*See D. E. Osterbrock and G. J. Ferland, *Astrophysics of Gaseous Nebulae and Active Galactic Nuclei* (University Science Books, Sausalito, 2006), chap. 2.

states except the ground energy level, which would just recycle the ionizing photon) $\alpha_B = 2.6 \times 10^{-13}$ cm^3 s^{-1} for hydrogen at $T = 10^4$ K. The complete description of the evolution of an expanding H II region, including the ingredients of a nonsteady ionizing source, recombinations, and cosmological expansion, is given by[53]

$$\bar{n}_{\mathrm{H}} \left(\frac{dV_p}{dt} - 3HV_p \right) = \frac{dN_\gamma}{dt} - \alpha_B \left\langle n_{\mathrm{H}}^2 \right\rangle V_p.$$

(5.3)

In this equation, the mean density $\bar{n}_{\mathrm{H}}$ varies with time as $1/a^3(t)$. Note that the recombination rate scales as the square of the density. Therefore, if the IGM is not uniform, but instead the gas which is being ionized is mostly distributed in high-density clumps, then the recombination time will be shorter. This is often accommodated for by introducing a volume-averaged clumping factor C (which is, in general, time dependent), defined by*

$$C = \left\langle n_{\mathrm{H}}^2 \right\rangle / \bar{n}_{\mathrm{H}}^2.$$

(5.4)

If the ionized volume is large compared to the typical scale of clumping, so that many clumps are averaged over, then equation (5.3) can be solved by supplementing it with equation (5.4) and specifying C. Switching to the

---

*The recombination rate depends on the number density of electrons, and in using equation (5.4) we are neglecting the small contribution made by partially or fully ionized helium.

comoving volume V, the resulting equation is

$$\frac{dV}{dt} = \frac{1}{\bar{n}_H^0} \frac{dN_\gamma}{dt} - \alpha_B \frac{C}{a^3} \bar{n}_H^0 V, \qquad (5.5)$$

where the present number density of hydrogen is

$$\bar{n}_H^0 = 2.1 \times 10^{-7} \text{ cm}^{-3}. \qquad (5.6)$$

This number density is lower than the total number density of baryons $\bar{n}_b^0$ by a factor of ~ 0.76, corresponding to the primordial mass fraction of hydrogen. The solution for $V(t)$ around a source which turns on at $t = t_i$ is[54]

$$V(t) = \int_{t_i}^t \frac{1}{\bar{n}_H^0} \frac{dN_\gamma}{dt'} e^{F(t',t)} dt', \qquad (5.7)$$

where

$$F(t', t) = -\alpha_B \bar{n}_H^0 \int_{t'}^t \frac{C(t'')}{a^3(t'')} dt''. \qquad (5.8)$$

At high redshifts $(z \gg 1)$, the scale factor varies as

$$a(t) \simeq \left(\frac{3}{2} \sqrt{\Omega_m} H_0 t \right)^{2/3}, \qquad (5.9)$$

and with the additional assumption of a constant C, the function F simplifies as follows. Defining

$$f(t) = a(t)^{-3/2}, \qquad (5.10)$$

we derive

$$F(t', t) = -\frac{2}{3} \frac{\alpha_B \bar{n}_{\mathrm{H}}^0}{\sqrt{\Omega_m} H_0} C[f(t') - f(t)]$$

$$= -0.26 \left(\frac{C}{10}\right) [f(t') - f(t)]. \quad (5.11)$$

The size of the resulting H II region depends on the halo that produces it. Let us consider a halo of total mass M and baryon fraction Ω_b/Ω_m. To derive a rough estimate, we assume that baryons are incorporated into stars with an efficiency of $f_\star = 10\%$, and that the escape fraction for the resulting ionizing radiation is also $f_{\mathrm{esc}} = 10\%$. As mentioned in chapter 4, Population II stars produce a total of $N_\gamma \approx 4{,}000$ ionizing photons per baryon in them. We may define a parameter which gives the overall number of ionizations per baryon,

$$N_{\mathrm{ion}} \equiv N_\gamma \, f_\star \, f_{\mathrm{esc}}. \quad (5.12)$$

If we neglect recombinations then we obtain the maximum comoving radius of the region which the halo of mass M can ionize as

$$r_{\max} = \left(\frac{3}{4\pi} \frac{N_\gamma}{\bar{n}_{\mathrm{H}}^0}\right)^{1/3} = \left(\frac{3}{4\pi} \frac{N_{\mathrm{ion}}}{\bar{n}_{\mathrm{H}}^0} \frac{\Omega_b}{\Omega_m} \frac{M}{m_p}\right)^{1/3}$$

$$= (680 \,\mathrm{kpc}) \left(\frac{N_{\mathrm{ion}}}{40} \frac{M}{10^8 M_\odot}\right)^{1/3}. \quad (5.13)$$

However, the actual radius never reaches this size if the recombination time is shorter than the lifetime of the ionizing source.

We may obtain a similar estimate for the size of the H II region around a galaxy if we consider a quasar rather than stars. For the typical quasar spectrum, $\sim 10^4$ ionizing photons are produced per baryon incorporated into the black hole, assuming a radiative efficiency of $\sim 6\%$. The overall efficiency of incorporating the collapsed fraction of baryons into the central black hole is low ($< 0.01\%$ in the local Universe), but $f_{\rm esc}$ is likely to be close to unity for powerful quasars which ionize their host galaxy. If we take the limit of an extremely bright source, characterized by an arbitrarily high production rate of ionizing photons, then equation (5.3) would imply that the H II region expands faster than light. This result is clearly unphysical and must be corrected for bright sources. At early times, the ionization front around a bright quasar would have expanded at nearly the speed of light, c, but this only occurs when the H II region is sufficiently small such that the production rate of ionizing photons by the central source exceeds their consumption rate by hydrogen atoms within this volume. It is straightforward to do the accounting for these rates (including recombination) by taking the light propagation delay into account. The general equation for the relativistic expansion of the *comoving* radius $r = (1 + z)r_p$ of a quasar H II region in an IGM with neutral filling fraction $x_{\rm HI}$ (fixed by other ionizing sources) is given by[55]

$$\frac{dr}{dt} = c(1 + z)$$

$$\times \left(\frac{\dot{N}_\gamma - \alpha_B C x_{\rm HI} \left(\bar{n}_{\rm H}^0 \right)^2 (1 + z)^3 \left(\frac{4\pi}{3} r^3 \right)}{\dot{N}_\gamma + 4\pi r^2 (1 + z) c x_{\rm HI} \bar{n}_{\rm H}^0} \right),$$

(5.14)

where $\dot{N}_\gamma$ is the rate of ionizing photons crossing a shell of the H II region at radius r and time t. Indeed, for $\dot{N}_\gamma \to \infty$ the propagation speed of the physical radius of the H II region $r_p = r/(1 + z)$ approaches the speed of light in the above expression, $(dr_p/dt) \to c$.

The process of the reionization of hydrogen involves several distinct stages.[56] The initial "preoverlap" stage consists of individual ionizing sources turning on and ionizing their surroundings. The first galaxies form in the most massive halos at high redshift, which are preferentially located in the highest-density regions. Thus, the ionizing photons that escape from the galaxy itself must then make their way through the surrounding high-density regions, characterized by a high recombination rate. Once they emerge, the ionization fronts propagate more easily through the low-density voids, leaving behind pockets of neutral, high-density gas. During this period, the IGM is a two-phase medium characterized by highly ionized regions separated from neutral regions by ionization fronts. Furthermore, the ionizing intensity is very inhomogeneous even within the ionized regions.

The central, relatively rapid "overlap" phase of reionization begins when neighboring H II regions begin to overlap. Whenever two ionized bubbles are joined, each point inside their common boundary becomes exposed to ionizing photons from both sources. Therefore, the ionizing intensity inside H II regions rises rapidly, allowing those regions to expand into high-density gas which had previously recombined fast enough to remain neutral when the ionizing intensity had been low. Since each bubble coalescence accelerates the process of reionization, the overlap phase has the character of a phase transition and is expected to occur rapidly. By the end of this stage, most regions in the IGM are able to "see" several unobscured sources, therefore, the ionizing intensity is much higher and more homogeneous than before overlap. An additional attribute of this rapid "overlap" phase results from the fact that hierarchical structure formation models predict a galaxy formation rate that rises rapidly with time at these high redshifts. This is because most galaxies fall on the exponential tail of the Press-Schechter mass function at early cosmic times, and so the fraction of mass that gets incorporated into stars grows exponentially with time. This process leads to a final state in which the low-density IGM has been highly ionized, with ionizing radiation reaching everywhere except gas located inside self-shielded, high-density clouds. This "moment of reionization" marks the end of the overlap phase.

Some neutral gas does, however, remain in high-density structures, which are gradually ionized as galaxy formation proceeds and the mean ionizing intensity grows with time. The ionizing intensity continues to grow and become more uniform as an increasing number of ionizing sources is visible to every point in the IGM.

Analytic models of the preoverlap stage focus on the evolution of the H II filling factor, i.e., the fraction of the volume of the Universe which is filled by H II regions, Q_{HII}. The modeling of individual H II regions can be used to understand the development of the total filling factor. Starting with equation (5.5), if we assume a common clumping factor C for all H II regions, then we can sum each term of the equation over all bubbles in a given large volume of the Universe and then divide by this volume. Then V can be replaced by the filling factor and N_γ by the total number of ionizing photons produced up to some time t, per unit volume. The latter quantity $\bar{n}_\gamma$ equals the mean number of ionizing photons per baryon multiplied by the mean density of baryons $\bar{n}_b$. Following the arguments leading to equation (5.13), we find that if we include only stars

$$\frac{\bar{n}_\gamma}{\bar{n}_b} = N_{\text{ion}} F_{\text{col}}, \qquad (5.15)$$

where the collapse fraction F_{col} is the fraction of all the baryons in the Universe which are in galaxies, i.e., the fraction of gas that has settled into halos and cooled efficiently inside them. In writing equation (5.15) we are assuming instantaneous production of photons, i.e., that the time scale for the formation and evolution of the massive stars in a galaxy is relatively short compared to the Hubble time at the formation redshift of the galaxy. The total number of ionizations equals the total number of ionizing photons produced by stars, i.e., all ionizing photons contribute regardless of the spatial distribution of sources. Also, the total recombination rate is proportional to the total ionized volume, regardless of its topology. Thus, even if two or more bubbles overlap, the model remains a good first approximation for Q_{HII} (at least until its value approaches unity).

Under these assumptions we convert equation (5.5), which describes individual H II regions, to an equation which statistically describes the transition from a neutral Universe to a fully ionized one:

$$\frac{d\,Q_{\text{HII}}}{dt} = \frac{N_{\text{ion}}}{0.76}\frac{d\,F_{\text{col}}}{dt} - \alpha_B\frac{C}{a^3}\bar{n}_{\text{H}}^0\,Q_{\text{HII}}, \qquad (5.16)$$

which admits the solution [in analogy with equation (5.7)],

$$Q_{\text{HII}}(t) = \int_0^t \frac{N_{\text{ion}}}{0.76}\frac{d\,F_{\text{col}}}{dt'}\,e^{F(t',t)}dt', \qquad (5.17)$$

where $F(t', t)$ is determined by equation (5.11).

A simple estimate of the collapse fraction at high redshift is the mass fraction [given by equation (3.17) in the Press-Schechter model] in halos above the cooling threshold, which gives the minimum mass of halos in which gas can cool efficiently. Assuming that only atomic cooling is effective during the redshift range of reionization, the minimum mass corresponds roughly to a halo of virial temperature $T_{\text{vir}} = 10^4\,\text{K}$, which can be converted to a mass using equation (3.12).

Although many models yield a reionization redshift around $z \sim 10$, the exact value depends on a number of uncertain parameters affecting both the source term and the recombination term in equation (5.16). The source parameters include the formation efficiency of stars and quasars and the escape fraction of ionizing photons produced by these sources.

The overlap of H II regions is expected to have occurred at different times in different regions of the IGM due to the cosmic scatter during the process of structure formation within finite spatial volumes. Reionization should be completed within a region of comoving radius R when the fraction of mass incorporated into collapsed objects in this region attains a certain critical value, corresponding to a threshold number of ionizing photons emitted per baryon. The ionization state of a region is governed by the factors of its enclosed ionizing luminosity, its overdensity, and dense pockets of neutral gas within the region that are self shielding to ionizing radiation. There is an offset δz between the redshift at which a region of mean overdensity $\bar{\delta}_R$ achieves this critical collapsed fraction, and the redshift $\bar{z}$ at which the Universe achieves the same collapsed fraction on average. This offset may be computed[57] from the expression for the collapsed fraction F_{col} within a region of overdensity $\bar{\delta}_R$ on a comoving scale R, within the excursion set formalism described in chapter 4, giving

$$F_{col} = \mathrm{erfc}\left(\frac{\delta_{crit} - \bar{\delta}_R}{\sqrt{2[\sigma_{R_{min}}^2 - \sigma_R^2]}} \right) \Rightarrow$$

$$\frac{\delta z}{(1 + \bar{z})} = \frac{\bar{\delta}_R}{\delta_{crit}(\bar{z})} - \left(1 - \sqrt{1 - \frac{\sigma_R^2}{\sigma_{R_{min}}^2}} \right), \quad (5.18)$$

where $\delta_{crit}(\bar{z}) \propto (1+\bar{z})$ is the collapse threshold for an overdensity at a redshift $\bar{z}$; and σ_R and $\sigma_{R_{min}}$ are the variances in the power spectrum linearly extrapolated to $z = 0$ on comoving scales corresponding to the region of interest and to the minimum galaxy mass M_{min}, respectively. The offset in the ionization redshift of a region depends on its linear overdensity $\bar{\delta}_R$. As a result, the distribution of offsets may be obtained directly from the power spectrum of primordial inhomogeneities. As can be seen from equation (5.18), larger regions have a smaller scatter due to their smaller cosmic variance. Interestingly, equation (5.18) is independent of the critical value of the collapsed fraction required for reionization.

The size distribution of ionized bubbles can also be calculated with an approximate analytic approach based on the excursion set formalism.[58] For a region to be ionized, galaxies inside it must produce a sufficient number of ionizing photons per baryon. This condition can be translated to the requirement that the collapsed fraction of mass in halos above some minimum mass $M_{\min}$ will exceed some threshold, namely $F_{\mathrm{col}} > \zeta^{-1}$. For example, requiring one ionizing photon per baryon would correspond to setting $\zeta = N_{\mathrm{ion}}$, where N_{ion} is defined in equation (5.12). We would like to find the largest region around every point whose collapse fraction satisfies the above condition on the collapse fraction, then calculate the abundance of ionized regions of this size. Different regions have different values of F_{col} because their mean density is different. Writing the collapse fraction in a region of mean overdensity δ_R as the left-hand side of equation (5.18), we may derive the barrier condition on the mean overdensity within a region of mass $M = \frac{4\pi}{3} R^3 \rho_m$ in order for it to be ionized,

$$\delta_R > \delta_B(M, z) \equiv \delta_c - \sqrt{2} K(\zeta) [\sigma_{\min}^2 - \sigma^2(M, z)]^{1/2},$$
$$(5.19)$$

where $K(\zeta) = \text{erfc}^{-1}(1 - \zeta^{-1})$. The barrier in equation (5.19) is well approximated by a linear dependence on σ^2 of the form $\delta_B \approx B(M) = B_0 + B_1\sigma^2(M)$, in which case the mass function has an analytic solution,

$$
\frac{dn}{dM} = \sqrt{\frac{2}{\pi}} \frac{\rho_m}{M^2} \left| \frac{d\ln\sigma}{d\ln M} \right| \frac{B_0}{\sigma(M)} \exp\left(-\frac{B^2(M)}{2\sigma^2(M)} \right),
$$

(5.20)

where ρ_m is the mean comoving mass density. This solution for $(dn/dM)dM$ provides the comoving number density of ionized bubbles with IGM mass in the range between M and $M + dM$. The main difference between this result and the Press-Schechter mass function is that the barrier in this case becomes more difficult to cross on smaller scales because δ_B is a decreasing function of mass M. This gives bubbles a characteristic size. The size evolves with redshift in a way that depends only on ζ and M_{min}.

A limitation of the above analytic model is that it ignores the nonlocal influence of sources on distant regions (such as voids) as well as the possible shadowing effect of intervening gas. Radiative transfer effects in the real Universe are inherently three dimensional and cannot be fully captured by spherical averages as done in this model. Moreover, the value of M_{min} is expected to increase in regions that were already ionized, complicating the expectation of whether they will remain ionized later. Nevertheless, refined versions of this model agree well with rigorous radiative transfer simulations.[59]

5.3 Swiss Cheese Topology

Detailed numerical simulations which evolve the formation of galaxies along with the radiative transfer of the ionizing photons they produce within a representative cosmological volume can provide a more accurate representation of the process of reionization than our approximate description above. The results from such simulations, illustrated in figure 5.1, demonstrate that the spatial distribution of ionized bubbles is indeed determined by clustered groups of galaxies and not by individual galaxies. At early times, galaxies were strongly clustered even on very large scales (up to tens of Mpc), and these scales therefore dominate the structure of reionization. The basic idea is simple:[60] at high redshift, galactic halos are rare and correspond to high-density peaks. As an analogy, imagine searching on Earth for mountain peaks above 5,000 meters. The 200 such peaks are not at all distributed uniformly, but instead are found in a few distinct clusters on top of large mountain ranges. Given the large-scale boost provided by a mountain range, a small-scale crest need only provide a small additional rise in order to become a 5,000 meter peak. The same crest, if it formed within a valley, would not come anywhere near 5,000 meters in total height. Similarly, in order to find the early galaxies, one should look in regions with large-scale density enhancements where galaxies are found in abundance.

The ionizing radiation emitted by the stars in each galaxy initially produces an isolated bubble of ionized gas. However, in a region dense with galaxies, the bubbles

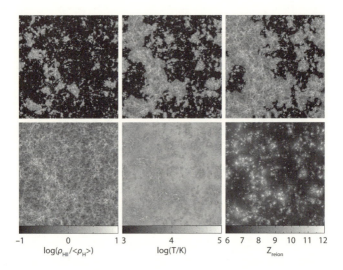

-1   0   1 3   4   5 6 7 8 9 10 11 12
$\log(\rho_{\mathrm{HII}}/<\rho_{\mathrm{H}}>)$ $\log(T/K)$ Z_{reion}

Figure 5.1. Snapshots from a numerical simulation illustrating the spatial structure of cosmic reionization in a slice of 140 comoving Mpc on a side. The simulation describes the dynamics of the dark matter and gas as well as the radiative transfer of ionizing radiation from galaxies. The first four panels (reading across from top left to bottom left) show the evolution of the ionized hydrogen density ρ_{HII} normalized by the mean proton density in the IGM $\langle \rho_{\mathrm{H}} \rangle = 0.76\Omega_b\bar{\rho}$ when the simulation volume is 25%, 50%, 75%, and 100% ionized, respectively. Large-scale overdense regions form large concentrations of galaxies whose ionizing photons produce joint ionized bubbles. At the same time, galaxies are rare within large-scale voids in which the IGM is mostly neutral at early times. The bottom middle panel shows the temperature at the end of reionization while the bottom right panel shows the redshift at which different gas elements are reionized. Higher-density regions tracing the large-scale structure are generally reionized earlier than lower-density regions far from sources. At the end of reionization, regions that were last to get ionized and heated are still typically hotter because they have not yet had time to cool through the cosmic expansion. The resulting inhomogeneities in the temperature of the IGM introduce spatial

quickly overlap into one large bubble, completing reionization in this region even while the rest of the universe is still mostly neutral. Most importantly, since the abundance of rare density peaks is very sensitive to small changes in the density threshold, even a large-scale region with a small density enhancement (say, 10% above the mean density of the Universe) can have a much larger concentration of galaxies than in other regions (characterized, for example, by a 50% enhancement). On the other hand, reionization is more difficult to achieve in dense regions since the protons and electrons collide and recombine more often in such regions, and newly formed hydrogen atoms need to be reionized again by additional ionizing photons. However, the overdense regions still end up reionizing first since the increase in the number of ionizing sources in these regions outweighs the higher recombination rate. The large-scale topology of reionization is therefore *inside out,* with underdense voids reionizing only at the very end of reionization using the help of extra ionizing photons coming in from their surroundings (which have a higher density of galaxies than the voids themselves). This is a key theoretical prediction awaiting observational testing.

Detailed analytical models accounting for large-scale variations in the abundance of galaxies confirm that the

variations in the cosmological Jeans mass, which in turn modulate the distribution of small galaxies [D. Babich and A. Loeb, *Astrophys. J.* **640**, 1 (2006)] and the Lyman-α forest [R. Cen, P. McDonald, H. Trac, and A. Loeb, *Astrophys. J.* **706**, L164 (2009)] at lower redshifts. Figure credit: H. Trac, R. Cen, and A. Loeb, *Astrophys. J.* **689**, L81 (2009).

typical bubble size starts well below a Mpc early in reionization, as expected for an individual galaxy, rises to 5–
10 Mpc during the central phase (i.e., when the Universe
is half ionized), and finally increases by another order of
magnitude toward the end of reionization. These scales
are given in comoving units that scale with the expansion
of the universe such that the actual sizes at a redshift z
were smaller than these numbers by a factor of $(1 + z)$.
Numerical simulations have only recently begun to reach
the enormous scales needed to capture this evolution.
Accounting precisely for gravitational evolution on a wide
range of scales, but still crudely for gas dynamics, star
formation, and the radiative transfer of ionizing photons,
the simulations confirm that the large-scale topology of
reionization is inside out, and that this "swiss cheese"
topology can be used to study the abundance and clustering of the ionizing sources (figure 5.1).

The characteristic observable size of the ionized bubbles
at the end of reionization can be calculated based on simple
considerations that depend only on the power spectrum
of density fluctuations and the redshift. As the size of
an ionized bubble increases, the time it takes a photon
emitted by hydrogen to traverse it gets longer. At the
same time, the variation in the time at which different
regions reionize becomes smaller as the regions grow larger.
Thus, there is a maximum region size above which the
photon-crossing time is longer than the cosmic variance
in reionization time. Regions larger than this size will be
ionized at their near side by the time a photon crosses
them toward the observer from their far side. They would
appear to the observer as one sided, and hence signal

the end of reionization. Using $\sigma(M)$ in figure 3.1, these considerations imply[61] a characteristic size for the ionized bubbles of ~ 10 physical Mpc at $z \sim 6$ (equivalent to 70 Mpc today). This result implies that future observations of the ionized bubbles (such as the low-frequency radio experiments described in chapter 6) should be tuned to a characteristic angular scale of tens of arcminutes for an optimal detection of brightness fluctuations in the emission of hydrogen near the end of reionization.

6

OBSERVING THE FIRST GALAXIES

6.1 Theories and Observations

A couple of years ago, my family and I visited the remote island of Tasmania off the coast of Australia, known for its unspoiled natural environment. Upon our arrival at a secluded lodge near the beautiful Cradle Mountain reserve, I discovered there was no internet connectivity. When night settled in, I had a few hours to spare—the time I ordinarily use to answer e-mails and check the daily postings of papers on the astro-ph web archive. I stepped out of our cabin and looked around at the pristine sight of nature left to its own. The night was dark with no city light anywhere on the horizon. Up on the sky was a magnificent view of the Milky Way galaxy stretched out in its full glory on a black background. For hours, I stared at our Galaxy's stars, dust, globular clusters, Magellanic clouds, as well as its sister galaxy—Andromeda, realizing at a deeper level that what we astronomers talk about truly exists out there. Eighteen months later, after he heard this story in a colloquium I gave about the Milky Way at Caltech, the distinguished astronomer Shri Kulkarni took me to an observing run on the 200-inch telescope at Mt. Palomar in a bold attempt to turn me into an observer.

There is wasted time in doing observations: the sky may be cloudy, nature may not cooperate in providing interesting results even if the observing campaign is successful, and there is considerable dead time in developing new instruments. Because of practical constraints associated with hardware, a typical observer has to develop a long-term program and is often restricted from changing directions on a short time scale. Theorists, on the other hand, can control the effective use of their research time and are free to switch quickly between topics if an exciting idea comes along from an unexpected direction. However, a theorist might spend a lifetime on a topic that turns out to be irrelevant as a description of reality, despite its being intellectually stimulating. From that perspective, *doing observations is the most efficient way of finding the truth*. But theoretical ideas are instrumental in guiding observers, since scientists occasionally miss what they are not set up to find.

6.2 Completing Our Photo Album of the Universe

The study of the first galaxies has so far been theoretical, but it is soon to become an observational frontier. How the primordial cosmic gas was reionized is one of the most exciting questions in cosmology today. Most theorists associate reionization with the first generation of stars, whose ultraviolet radiation streamed into intergalactic space and broke hydrogen atoms apart. Others conjecture that material plummeting into an unknown population of low-mass black holes gave off sufficient radiation on its death plunge. New observational data are required to test which of these scenarios better describes reality. The timing of reionization depends on astrophysical parameters such as the efficiency of making stars or black holes in galaxies.

Let us summarize quickly what we have learned in the previous chapters. According to the popular cosmological model of cold dark matter, dwarf galaxies started to form when the Universe was only a hundred million years old. Computer simulations indicate that the first stars to have formed out of the primordial gas left over from the Big Bang were much more massive than the Sun. Lacking heavy elements to cool the gas to lower temperatures, the warm primordial gas could have only fragmented into relatively massive clumps which condensed to make the first stars. These stars were efficient factories of ionizing radiation. Once they exhausted their nuclear fuel, some of these stars exploded as supernovae and dispersed the heavy elements cooked by nuclear reactions in their interiors into the surrounding gas. The heavy elements cooled the diffuse gas to lower temperatures and allowed it to fragment into lower-mass clumps that made the second generation of stars. The ultraviolet radiation emitted by all genera-tions of stars eventually leaked into the intergalactic space and ionized gas far outside the boundaries of individual galaxies.

The earliest dwarf galaxies merged and made bigger galaxies as time went on. A present-day galaxy like our own Milky Way was constructed over cosmic history by the assembly of a million building blocks in the form of the first dwarf galaxies. The UV radiation from each galaxy created an ionized bubble in the cosmic gas around it. As the galaxies grew in mass, these bubbles expanded in size and eventually surrounded whole groups of galaxies. Finally, as more galaxies formed, the bubbles overlapped and the initially neutral gas in between the galaxies was completely reionized.

Although the above progression of events sounds plausible, at this time it is only a thought floating in the minds of theorists that has not yet received confirmation from observational data. Empirical cosmologists would like to actually see direct evidence for the reionization process before accepting it as common knowledge. *How can one observe the reionization history of the Universe directly?*

One way is to search for the radiation emitted by the first galaxies using large new telescopes from the ground as well as from space. Another way is to image hydrogen and study the cavities of ionized bubbles within it. Here and in chapter 7 we will describe each of these techniques. The observational exploration of the reionization epoch promises to be one of the most active frontiers in cosmology over the coming decade.

6.3 Cosmic Time Machine

When we look at our image reflected off a mirror at a distance of 1 meter, we see the way we looked 6 nanoseconds ago, the time it took light to travel to the mirror and back. If the mirror is spaced 10^{19} cm $= 3$ pc away, we will see the way we looked 21 years ago. Light propagates at a finite speed, so by observing distant regions, we are able to see how the Universe looked in the past, a light travel time ago (see figure 2.2). The statistical homogeneity of the Universe on large scales guarantees that what we see far away is a fair statistical representation of the conditions that were present in our region of the Universe a long time ago.

This fortunate situation makes cosmology an empirical science. We do not need to guess how the Universe evolved. By using telescopes we can simply see the way distant regions appeared at earlier cosmic times. Since a greater distance means a fainter flux from a source of a fixed luminosity, the observation of the earliest sources of light requires the development of sensitive instruments, and poses technological challenges to observers.

We can image the Universe only if it is transparent. Earlier than 400 thousand years after the Big Bang, the cosmic gas was sufficiently hot to be fully ionized (i.e., atoms were broken into free nuclei and electrons), and the Universe was opaque due to scattering by the dense fog of free electrons that filled it. Thus, telescopes cannot be used to image the infant Universe at earlier times (at redshifts $>10^3$). The earliest possible image of the Universe can be seen in the cosmic microwave background, the thermal radiation left over from the transition to transparency (figure 1.1).

How will the earliest galaxies appear to our telescopes? We can easily express the flux observed from a galaxy of luminosity L at a redshift z. The observed flux (energy per unit time per unit telescope area) is obtained by spreading the energy emitted from the source per unit time, L, over the surface area of a sphere whose radius equals to the effective distance of the source,

$$f = \frac{L}{4\pi d_L^2}, \tag{6.1}$$

where d_L is defined as the *luminosity distance* in cosmology. For a flat Universe, the comoving distance of a galaxy which emitted its photons at a time t_{em} and is observed at time t_{obs} is obtained by summing over infinitesimal distance elements along the path length of a photon, $c\,dt$, each expanded by a factor $(1 + z)$ to the present time:

$$r_{em} = \int_{t_{em}}^{t_{obs}} \frac{c\,dt}{a(t)} = \frac{c}{H_0} \int_0^z \frac{dz'}{\sqrt{\Omega_m(1 + z')^3 + \Omega_\Lambda}}, \tag{6.2}$$

where $a = (1 + z)^{-1}$. The *angular diameter distance* d_A, corresponding to the angular diameter $\theta = D/d_A$ occupied by a galaxy of size D, must take into account the fact that we were closer to that galaxy* by a factor $(1 + z)$ when the photons started their journey at a redshift z, so it is simply given by $d_A = r_{em}/(1 + z)$. But to find d_L we must take account of additional redshift factors.

If a galaxy has an intrinsic luminosity L, then it will emit an energy $L dt_{em}$ over a time interval dt_{em}. This energy is redshifted by a factor of $(1 + z)$ and is observed over a longer time interval $dt_{obs} = dt_{em}(1 + z)$ after

---

*In a flat Universe, photons travel along straight lines. The angle at which a photon is seen is not modified by the cosmic expansion, since the Universe expands at the same rate both parallel and perpendicular to the line of sight.

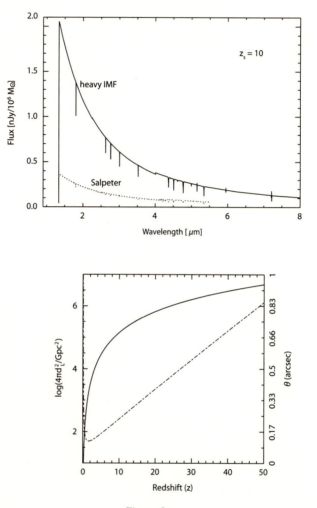

Figure 6.1.

being spread over a sphere of surface area $4\pi r_{\mathrm{em}}^2$. Thus, the observed flux will be

$$f = \frac{L \, dt_{\mathrm{em}}/(1+z)}{4\pi r_{\mathrm{em}}^2 \, dt_{\mathrm{obs}}} = \frac{L}{4\pi r_{\mathrm{em}}^2 (1+z)^2}, \qquad (6.3)$$

implying that[*]

$$d_L = r_{\mathrm{em}}(1+z) = d_A(1+z)^2. \qquad (6.4)$$

The area dilution factor $4\pi d_L^2$ is plotted as a function of redshift in the bottom panel of figure 6.1. If the

---

[*]A simple analytic fitting formula for $d_L(z)$ was derived by U.-L. Pen, *Astrophys. J. Suppl.* **120**, 49 (1999); http://arxiv.org/pdf/astro-ph/9904172v1.

Figure 6.1. *Facing page, top:* Comparison of the observed flux per unit frequency from a galaxy at a redshift $z_s = 10$ for a Salpeter IMF [*dotted line*; J. Tumlinson and M. J. Shull, *Astrophys. J.* **528**, L65 (2000)] and a purely massive IMF [*solid line*; V. Bromm, R. P. Kudritzki, and A. Loeb, *Astrophys. J.* **552**, 464 (2001)]. The flux in units of nJy per $10^6 M_\odot$ of stars is plotted as a function of observed wavelength in μm. The cutoff below an observed wavelength of $(1216 \text{Å})(1 + z_s) = 1.34 \mu$m is due to hydrogen Lyman-α absorption in the IGM (the so-called Gunn-Peterson effect; see chapter 7). For the same total stellar mass, the observable flux is larger by an order of magnitude for stars biased toward having masses $>100 M_\odot$. *Bottom:* The solid line (corresponding to the label on the left-hand side) shows $\log_{10}$ of the conversion factor between the luminosity of a source and its observed flux, $4\pi d_L^2$ (in Gpc^2), as a function of redshift z. The dash-dotted line (labeled on the right) gives the angle θ (in arcseconds) occupied by a galaxy of a 1 kpc diameter as a function of redshift.

observed flux is measured over only a narrow band of frequencies, one needs to take account of the additional conversion factor of $(1 + z) = (d\nu_{em}/d\nu_{obs})$ between the emitted frequency interval $d\nu_{em}$ and its observed value $d\nu_{obs}$. This yields the relation $(df/d\nu_{obs}) = (1 + z) \times (dL/d\nu_{em})/(4\pi d_L^2)$. The top panel of figure 6.1 compares the predicted flux per unit frequency* from a galaxy at a redshift $z_s = 10$ for a Salpeter IMF and for massive ($>100 M_\odot$) Population III stars, in units of nJy per $10^6 M_\odot$ in stars (where $1 \, \text{nJy} = 10^{-32} \, \text{erg} \, \text{cm}^{-2} \, \text{s}^{-1} \, \text{Hz}^{-1}$). The observed flux is an order of magnitude larger in the Population III case. The strong UV emission by massive stars is likely to produce bright recombination lines (such as Lyman-α and He II 1640 Å) from the interstellar medium surrounding these stars.

Theoretically, the expected number of early galaxies of different fluxes per unit area on the sky can be calculated by dressing up the dark matter halos in figure 3.2 with stars of some prescribed mass distribution and formation history, then finding the corresponding abundance of galaxies of different luminosities as a function of redshift.[62] There are many uncertain parameters in this approach (such as $f_\star$, f_{esc}, the stellar mass function, the star formation time, the metallicity, and feedback), so one is tempted to calibrate these parameters by observing the sky.[63]

---

*The observed flux per unit frequency can be translated to an equivalent *AB magnitude* using the relation AB $\equiv -2.5 \log_{10}[(df/d\nu_{obs})/\text{erg} \, \text{s}^{-1} \, \text{cm}^{-2} \, \text{Hz}^{-1}] - 48.6$

6.4 The Hubble Deep Field and Its Follow-Ups

In 1995, Bob Williams, then Director of the Space Telescope Science Institute, invited leading astronomers to advise him where to point the Hubble Space Telescope (HST) during the discretionary time he received as a Director, which amounted to a total of up to 10% of HST's observing time.* Each of the invited experts presented a detailed plan for using HST's time in sensible, but complex, observing programs addressing their personal research interests. After much of the day had passed, it became obvious that no consensus would be reached. "What shall we do?" asked one of the participants. Out of desperation, another participant suggested, "Why don't we point the telescope toward a fixed nonspecial direction and burn a hole in the sky as deep as we can go?"—just like checking how fast your new car can go. This simple compromise won the day since there was no real basis for choosing among the more specialized suggestions. As it turned out, this "hole burning" choice was one of the most influential uses of the HST as it produced the deepest image we have so far of the cosmos.

The Hubble Deep Field (HDF) covered an area of 5.3 squared arcminutes and was observed over 10 days (see figure 6.2). One of its pioneering findings was the discovery of large numbers of high-redshift galaxies at a time when only a small number of galaxies at $z > 1$

---

*E. Turner, private communication (2009).

Figure 6.2. The first Hubble Deep Field (HDF) image taken in 1995. The HDF covers an area 2.5 arcminutes across and contains a few thousand galaxies (with a few candidates up to a redshift $z \sim 6$). The image was taken in four broadband filters centered on wavelengths of 3000, 4500, 6060, and 8140 Å, with an average exposure time of ~ 0.127 million seconds per filter.

were known.* The HDF contained many red galaxies with some reaching a redshift as high as 6, or even higher.[64] The wealth of galaxies discovered at different stages of their

---

*Just prior to the HDF, an important paper about high-redshift galaxies was declined for publication because the referee pointed out the "well-known fact" that there are no galaxies beyond a redshift of 1.

evolution allowed astronomers to estimate the variation in the global rate of star formation per comoving volume over the lifetime of the universe.

Subsequent incarnations of this successful approach included the HDF-South and the Great Observatories Origins Deep Survey (GOODS). A section of GOODS, occupying a tenth of the diameter of the full moon (equivalent to 11 square arcminutes), was then observed for a total exposure time of a million seconds to create the Hubble Ultra Deep Field (HUDF), the most sensitive deep field image in visible light to date.* Red galaxies were identified in the HUDF image up to a redshift of $z \sim 7$, and possibly even higher, showing that the typical UV luminosity of galaxies declines with redshift at $z > 4$.[65] The redshifts of galaxies are inferred either through a search for a Lyman-α emission line (identifying so-called Lyman-α galaxies),[66] or through a search for a spectral break associated with the absorption of intervening hydrogen (so-called Lyman-break galaxies).[67] For very faint sources, redshifts are only identified crudely based on the spectral trough produced by hydrogen absorption in the host galaxy and the IGM (see chapter 7).

The abundance of Lyman-α galaxies shows a strong decline between $z = 5.7$ and $z = 7$, as expected from a correspondingly rapid increase in the neutral fraction of the IGM (which would scatter the Lyman-α line photons and make the line emission from these galaxies undetectable),[68] but this interpretation is not unique.

---

*In order for galaxy surveys to be statistically reliable, they need to cover large areas of the sky. Counts of galaxies in small fields of view suffer from a large cosmic variance owing to galaxy clustering.

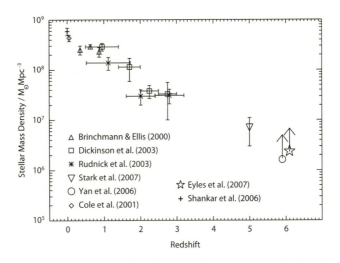

Figure 6.3. The observed evolution in the cosmic mass density of stars (in solar masses per comoving Mpc^3) as a function of redshift [data compiled by L. P. Eyles et al., *Mon. Not. R. Astron. Soc.* **374**, 910 (2007)]. The estimates at $z > 5$ should be regarded as lower limits due to the missing contribution of low-luminosity galaxies below the detection threshold [D. Stark et al., *Astrophys. J.* **659**, 84 (2007)]. Nevertheless, the data show that less than a few percent of all present-day stars had formed at $z > 5$, in the first 1.2 billion years after the Big Bang. A minimum density $\sim 1.7 \times 10^6 f_{\mathrm{esc}}^{-1} M_{\odot}$ Mpc^{-3} of Population II stars (corresponding to $\Omega_{\star} \sim 1.25 \times 10^{-5} f_{\mathrm{esc}}^{-1}$) is required to produce one ionizing photon per hydrogen atom in the Universe.

The mass budget of stars at $z \sim$ 5–6 has been inferred from complementary infrared observation with the Spitzer Space Telescope (see figure 6.3). The mean age of the stars in individual galaxies implies that they had formed at $z \sim 10$ and could have produced sufficient photons to reionize the IGM.

Another approach adopted by observers benefits from magnifying devices provided for free by nature, so-called gravitational lenses. Rich clusters of galaxies have such a large concentration of mass that their gravity bends the light rays from any source behind them and magnifies its image. This allows observers to probe fainter galaxies at higher redshifts than ever probed before. The redshift record from this method is currently associated[69] with a strongly lensed galaxy at $z = 7.6$. As of the writing of this book, this method has provided candidate galaxies with possible redshifts up to $z \sim 10$, but without further spectroscopic confirmation that would make these detections robust.[70]

So far, we have not seen the first generation of dwarf galaxies at redshifts $z > 10$ that were responsible for reionization.

6.5 Observing the First Gamma-Ray Bursts

Explosions of individual massive stars (such as supernovae) can also outshine their host galaxies for brief periods of time. The brightest among these explosions are *gamma-ray bursts* (GRBs), observed as short flashes of high-energy photons followed by afterglows at lower photon energies (as discussed in section 4.4). These afterglows can be used to study the first stars directly. Also, similarly to quasars, these beacons of light probe the state of the cosmic gas through its absorption line signatures on their spectra along the line of sight. GRBs were discovered by the Swift satellite out to a record redshift of $z = 8.3$, merely 620 million years after the Big Bang, and significantly earlier than the farthest known quasar ($z = 6.4$; see figure 7.2). It is already evident that GRB observations hold the promise of opening a new window into the infant Universe.

Standard light bulbs appear fainter with increasing redshift, but this is not the case with GRBs, which are transient events that fade with time. When observing a burst at a constant time delay, we are able to see the source at an earlier time in its own frame. This is a simple consequence of time stretching due to the cosmological redshift. Since the bursts are brighter at earlier times, it turns out that detecting them at high redshifts is almost as feasible as finding them at low redshifts, when they are closer to us.[71] It is a fortunate coincidence that the brightening associated with seeing the GRB at an intrinsically earlier time roughly compensates for the dimming associated with the increase in distance to the higher redshift.

In contrast to bright quasars, GRBs are expected to reside in typical small galaxies where massive stars form at those high redshifts. Once the transient GRB afterglow fades away, observers may search for the steady but weaker emission from its host galaxy. High-redshift GRBs may therefore serve as signposts of high-redshift galaxies which are otherwise too faint to be identified on their own. Also, in contrast to quasars, GRBs (and their faint host galaxies) have a negligible influence on the surrounding intergalactic medium. This is because the bright UV emission of a GRB lasts less than a day, compared with tens of millions of years for a quasar. Therefore, bright GRBs are unique in that they probe the true ionization state of the surrounding medium without modifying it.[72]

Figure 6.4. A full-scale model of the James Webb Space Telescope (JWST), the successor to the Hubble Space Telescope (http://www.jwst.nasa.gov/). JWST includes a primary mirror 6.5 meters in diameter, and offers instrument sensitivity across the infrared wavelength range of 0.6–28 μm, which will allow detection of the first galaxies. The size of the Sun shield (the large flat screen in the image) is 22 meters $\times$ 10 meters (72 ft $\times$ 29 ft). The telescope will orbit 1.5 million kilometers from Earth at the Lagrange L2 point.

As discussed in chapter 4, long-duration GRBs are believed to originate from the collapse of massive stars at the end of their lives (figure 4.5). Since the very first stars were likely massive, they could have produced GRBs.[73] If they did, we may be able to see them one star at a time. The discovery of a GRB afterglow whose spectroscopy indicates a metal-poor gaseous environment could potentially signal

Figure 6.5.

the first detection of a Population III star. Various space missions are currently proposed to discover GRBs at the highest possible redshifts.

6.6 Future Telescopes

The first stars emitted their radiation primarily in the UV band, but because of intergalactic absorption and their exceedingly high redshift, their detectable radiation is mostly observed in the infrared band. The successor to the Hubble Space Telescope, the James Webb Space Telescope (JWST), will include an aperture 6.5 meters in diameter, made of gold-coated beryllium and designed to operate in the infrared wavelength range of 0.6–28 μm (see figure 6.4). JWST will be positioned at the Lagrange L2 point, where any free-floating test object stays in the opposite direction to that of the Sun relative to Earth. The earliest galaxies are expected to be extremely faint and compact, for two reasons: first, they are associated with the smallest gaseous objects to have condensed out of the primordial gas, and second, they are located at the greatest distances from us among all galaxies.[74] Endowed with a large aperture and positioned outside the Earth's atmospheric emissions and opacity, JWST is ideally suited

Figure 6.5. *Facing page:* Artist's conception of the designs for three future giant telescopes that will be able to probe the first generation of galaxies from the ground: the European Extremely Large Telescope (EELT, top), the Giant Magellan Telescope (GMT, middle), and the Thirty Meter Telescope (TMT, bottom).

for resolving the faint glow from the first galaxies. It would be particularly exciting if JWST finds spectroscopic evidence for metal-free (Population III) stars. As shown in figure 6.1, the smoking gun signature would be a spectrum with no metal lines, a strong UV continuum consistent with a blackbody spectrum of $\sim 10^5$ K truncated by an IGM absorption trough (at wavelengths shorter than Lyman-α in the source frame; see section 7.2), and strong helium recombination lines (including a line at 1640 Å to which the IGM is transparent) from the interstellar gas around these hot stars.[75]

Several initiatives to construct large infrared telescopes on the ground are also under way. The next generation of ground-based telescopes will have an effective diameter of 24–42 meters; examples include the European Extremely Large Telescope,[76] the Giant Magellan Telescope,[77] and the Thirty Meter Telescope,[78] which are illustrated in figure 6.5. Along with JWST, they will be able to image and survey a large sample of early galaxies. Given that these galaxies also created ionized bubbles during reionization, their locations should be correlated with the existence of cavities in the distribution of neutral hydrogen. Within the next decade it may become feasible to explore the environmental influence of galaxies by using infrared telescopes in concert with radio observatories that will map diffuse hydrogen at the same redshifts[79] (see section 7.3). Additional emission at submillimeter wavelengths from molecules (such as CO), ions (such as C II), atoms (such as O I), and dust within the first galaxies would potentially be detectable with the future Atacama Large Millimeter/Submillimeter Array (ALMA).[80]

What makes the study of the first galaxies so exciting is that it involves work in progress. If all the problems were solved, there would be nothing left to be discovered by future scientists, such as some of the young readers of this book. Scientific knowledge often advances like a burning front, in which the flame is more exciting than the ashes. It would obviously be rewarding if our current theoretical ideas are confirmed by future observations, but it might even be more exciting if these ideas are modified.

I started to work on the first galaxies two decades ago when there were only a few theorists around the globe interested in this field of research. It is gratifying to see an explosive evolution of this frontier now, with the further potential of a wealth of new instruments under construction. Astronomers are looking forward to observing through these new telescopes and discovering, for the first time in human history, how the very first galaxies and stars formed.

7

IMAGING THE DIFFUSE FOG OF COSMIC HYDROGEN

7.1 Hydrogen

Hydrogen is the most abundant element in the Universe. It is also the simplest atom possible, containing a proton and an electron held together by their mutual electric attraction. Because of its simplicity, the detailed understanding of the hydrogen atom structure played an important role in the development of quantum mechanics.

Since the lifetime of energy levels with principal quantum number n greater than 1 is far shorter than the typical time it takes to excite them in the rarefied environments of the Universe, hydrogen is commonly found to be in its ground state (lowest energy level) with $n = 1$. This implies that the transitions we should focus on are those that involve the $n = 1$ state. Below we describe two such transitions, depicted in figure 7.1.

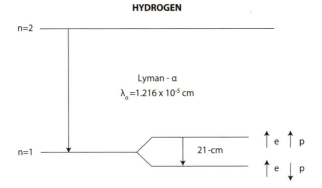

Figure 7.1. Two important transitions of the hydrogen atom. The 21-cm transition of hydrogen is between two slightly separated (hyperfine) states of the ground energy level (principal quantum number $n = 1$). In the higher-energy state, the spin of the electron (e) is aligned with that of the proton (p), and in the lower-energy state the two are antialigned. A spin flip of the electron results in the emission of a photon with a wavelength of 21 cm (or a frequency of 1420 MHz). The second transition is between the $n = 2$ and the $n = 1$ levels, resulting in the emission of a Lyman-α photon of wavelength $\lambda_\alpha = 1.216 \times 10^{-5}$ cm (or a frequency of 2.468×10^{15} Hz).

7.2 The Lyman-α Line

The most widely discussed transition of hydrogen in cosmology is the Lyman-α spectral line, which was discovered experimentally in 1905 by Harvard physicist Theodore Lyman. This line has been traditionally used to probe the ionization state of the IGM in the spectra of quasars, galaxies, and gamma-ray bursts. Back in 1965, Peter Scheuer[81] and, independently, Jim Gunn and Bruce Peterson[82] realized that the cross section for Lyman-α absorption is so large that the IGM should be opaque to it[*] even if its neutral (nonionized) fraction is as small as $\sim 10^{-5}$. The lack of complete absorption for quasars at redshifts $z < 6.4$ is now interpreted as evidence that the diffuse IGM was fully ionized within less than a billion years after the Big Bang. Quasar spectra do show evidence for a so-called forest of Lyman-α absorption features, which originate from slight enhancements in the tiny fraction of hydrogen within overdense regions of the cosmic web. The Lyman-α forest has so far been observed in the spectrum of widely separated "skewers" pointing toward individual quasars. Since the cosmic web provides a measure of the power spectrum $P(k)$, there are plans to observe a dense array of skewers associated with a large number of quasars and map the related large-scale structure it delineates in three dimensions.

If a source were to be observed before or during the epoch of reionization, when the atomic fraction of hydrogen was more substantial, then all photons with wavelengths just short of the Lyman-α wavelength at the source [observed at $\lambda_\alpha = 1216(1 + z_s)$ Å, where z_s is the source redshift] would redshift into resonance, be absorbed by the IGM, and then get reemitted in other directions.[†] This would result in an observed absorption trough

---

[*]The optical depth for Lyman-α absorption out to a source at a redshift z_s by an IGM with an atomic hydrogen fraction x_{HI} is $\tau \approx 7 \times 10^5 x_{\mathrm{HI}}[(1 + z_s)/10]^{3/2}$.

[†]The scattering by the IGM generates a polarized Lyman-α halo around each source, which could also be used to measure the neutral fraction of the IGM; see A. Loeb and G. Rybicki, *Astrophys. J.* **524**, 527 (1999) and **520**, L79 (1999), and also M. Dijkstra and A. Loeb, *Mon. Not. R. Astron. Soc.* **386**, 492 (2008).

shortward of λ_α in the source spectrum, known as the "Gunn-Peterson effect." Often, quasars or gas-rich galaxies have a Lyman-α emission line, but the Gunn-Peterson effect is expected to eliminate the short-wavelength wing of the line (and potentially damp the entire Lyman-α emission feature if the line is sufficiently narrow).

The Gunn-Peterson trough serves as a robust indicator for the redshift of quasars, galaxies, and gamma-ray bursts during the epoch of reionization. Since it represents a broad spectral feature, its existence can be inferred by binning photons across broad frequency bands, a techniques labeled by astronomers as "photometry" (see figure 7.2). This approach is particularly handy for faint galaxies that supply a small number of photons during an observing run, since fine binning of their frequency distribution (commonly termed "spectroscopy") is impractical. In this context, the rarer bright sources have an important use. The Lyman-α cross section is so large that absorption extends to wavelengths even slightly longer than λ_α, creating the appearance of a smooth wing with a characteristic shape. A spectroscopic detection of the detailed shape of this Lyman-α damping wing can be used to infer the neutral fraction of hydrogen in the IGM during reionization.[83]

The spectra of the highest-redshift quasars at $z < 6.4$ show hints of a Gunn-Peterson effect (see figure 7.2), but the evidence is not conclusive since the observed high opacity of the Lyman-α transition can also result from trace amounts of hydrogen. A more fruitful approach would be to image hydrogen directly through a ground-state transition with a weaker opacity, a possibility we explore next.

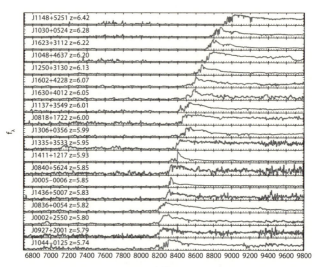

Figure 7.2. Observed spectra (flux per unit wavelength) of 19 quasars with redshifts $5.74 < z < 6.42$ from the Sloan Digital Sky Survey. For some of the highest-redshift quasars, the spectrum shows no transmitted flux shortward of the Lyman-α wavelength at the quasar redshift, providing a possible hint of the so-called Gunn-Peterson trough and indicating a slightly increased neutral fraction of the IGM. It is evident from these spectra that broadband photometry is adequate for inferring the redshift of sources during the epoch of reionization. Figure credit: X. Fan et al., *Astron. J.* **128**, 515 (2004).

7.3 The 21-cm Line

In quantum mechanics, elementary particles possess a fundamental property called "spin" (classically thought of as the rotation of a particle around its axis), which has a half-integer or integer magnitude, and an up or down state.

The ground state of hydrogen is split into two very close ("hyperfine") states: an upper energy level (triplet state) in which the spin of the electron is lined up with that of the proton, and a lower energy level (singlet state) in which the two are antialigned. The transition to the lower level is accompanied by the emission of a photon with a wavelength of 21 centimeters (see figure 7.1). The 21-cm transition was theoretically predicted by Hendrick van de Hulst in 1944 and detected in 1951 by Harold Ewen and Ed Purcell, who put a horn antenna out of an office window in the Harvard physics department and saw the 21-cm emission from the Milky Way.

The existence of neutral hydrogen prior to reionization offers the prospect of detecting its 21-cm emission or absorption relative to the CMB.[84] The optical depth is only $\sim$1% in this case for a fully neutral IGM, making the 21-cm line a more suitable probe for the epoch of reionization than the Lyman-α line. By observing different wavelengths of $(21\,\text{cm}) \times (1 + z)$, one is slicing the Universe at different redshifts z. The redshifted 21-cm emission should display angular structure as well as frequency structure due to inhomogeneities in the gas density, the hydrogen ionized fraction, and the fraction of excited atoms. A full map of the distribution of atomic hydrogen (denoted by astronomers as H I) as a function of redshift would provide a three-dimensional image of the swiss-cheese structure of the IGM during reionization, as illustrated in figure 7.3. The cavities in the hydrogen distribution are the ionized bubbles around groups of early galaxies.

The relative population of the two levels defines the so-called *spin (excitation) temperature*, which may deviate

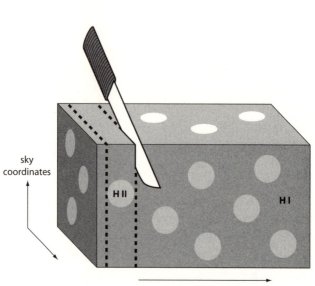

Figure 7.3. 21-cm imaging of ionized bubbles during the epoch of reionization, analogous to slicing swiss cheese. The technique of slicing at intervals separated by the typical dimension of a bubble is optimal for revealing different patterns in each slice.

from the ordinary (kinetic) temperature of the gas in the presence of a radiation field. The coupling between the gas and the microwave background (owing to the small residual fraction of free electrons left over from the hydrogen formation epoch) kept the gas temperature equal to the radiation temperature for up to 10 million years after the Big Bang. Subsequently, the cosmic expansion cooled the gas faster than the radiation, and collisions among the atoms maintained their spin temperature at equilibrium with their own kinetic temperature. At this phase, cosmic hydrogen can be detected in *absorption* against the microwave background sky since it is colder. Regions that are somewhat denser than the mean will produce more absorption and underdense regions will produce less absorption. The resulting fluctuations in the 21-cm brightness simply reflect the primordial inhomogeneities in the gas.[85] A hundred million years after the Big Bang, cosmic expansion diluted the density of the gas to the point where the collisional coupling of the spin temperature to the gas became weaker than its coupling to the microwave background. At this stage, the spin temperature returned to equilibrium with the radiation temperature, and it became impossible to see the gas against the microwave background brightness. Once the first galaxies lit up, they heated the gas (mainly by emitting X-rays which penetrated the thick column of intergalactic hydrogen) as well as its spin temperature (through UV photons which couple the spin temperature to the gas kinetic temperature). The increase of the spin temperature beyond the microwave background temperature requires much less energy per atom than ionization, so this heating occurred well before the Universe was reionized. Once the spin temperature had risen above the microwave background (CMB) temperature, the gas could be seen against the microwave sky in *emission*. At this stage, the hydrogen distribution is punctuated with bubbles of ionized gas which are created around groups of galaxies. Below we describe these evolutionary stages more quantitatively.

The basic physics of the hydrogen spin transition is determined as follows. The ground-state hyperfine levels of hydrogen tend to thermalize with the CMB, making the IGM unobservable. If other processes shift the hyperfine level populations away from thermal equilibrium, then the gas becomes observable against the CMB either in emission or in absorption. The relative occupancy of the spin levels is usually described in terms of the hydrogen spin temperature T_S, defined by

$$\frac{n_1}{n_0} = 3 \exp\left\{-\frac{T_*}{T_S}\right\}, \qquad (7.1)$$

where n_0 and n_1 refer respectively to the singlet and triplet hyperfine levels in the atomic ground state ($n = 1$), and $T_* = 0.068\,\mathrm{K}$ is defined by $k_B T_* = E_{21}$, where the energy of the 21-cm transition is $E_{21} = 5.9 \times 10^{-6}\,\mathrm{eV}$, corresponding to a photon frequency of 1,420 MHz. In the presence of the CMB alone, the spin states reach thermal equilibrium with $T_S = T_{\mathrm{CMB}} = 2.73(1 + z)\,\mathrm{K}$ on a time scale of $\sim T_*/(T_{\mathrm{CMB}} A_{10}) = 3 \times 10^5 (1 + z)^{-1}$ yr, where $A_{10} = 2.87 \times 10^{-15}\,\mathrm{s}^{-1}$ is the spontaneous decay rate of the hyperfine transition. This time scale is much shorter than the age of the Universe at all redshifts after cosmological recombination.

The IGM is observable only when the kinetic temperature T_k of the gas (defined by the motion of its atoms) differs from T_{CMB}, and an effective mechanism couples T_S to T_k. At early times, collisions dominate this coupling because the gas density is still high, but once a significant galaxy population forms in the Universe, the spin temperature is affected also by an indirect mechanism that acts through the scattering of Lyman-α photons, the so-called Wouthuysen-Field effect, named after the Dutch physicist Siegfried Wouthuysen and Harvard astrophysicist George Field, who explored it first.[86] Here, continuum UV photons produced by early radiation sources redshift by the Hubble expansion into the local Lyman-α line at a lower redshift and mix the spin states.

A patch of neutral hydrogen at the mean density and with a uniform T_S produces (after correcting for stimulated emission) an optical depth at an observed wavelength of $21(1 + z)$ cm of

$$\tau(z) = 1.1 \times 10^{-2} \left(\frac{T_{CMB}}{T_S} \right) \left(\frac{1 + z}{10} \right)^{1/2}, \quad (7.2)$$

where we have assumed $z \gg 1$. The observed spectral intensity I_ν relative to the CMB at a frequency ν is measured by radio astronomers as an effective brightness temperature T_b of blackbody emission at this frequency, defined using the Rayleigh-Jeans limit of the Planck formula, $I_\nu \equiv 2k_B T_b \nu^2 / c^2$.

The brightness temperature through the IGM is $T_b = T_{CMB} e^{-\tau} + T_S(1 - e^{-\tau})$, so the observed differential antenna temperature of this region relative to the CMB is[87]

$$T_b = (1 + z)^{-1}(T_S - T_{CMB})(1 - e^{-\tau})$$

$$\simeq (29 \text{ mK}) \left(\frac{1 + z}{10} \right)^{1/2} \left(\frac{T_S - T_{CMB}}{T_S} \right), \quad (7.3)$$

where we have made use of the fact that $\tau \ll 1$ and T_b has been redshifted to $z = 0$. The abbreviated unit mK stands for milliKelvin or 10^{-3} K.

In overdense regions, the observed T_b is proportional to the overdensity, while in partially ionized regions T_b is

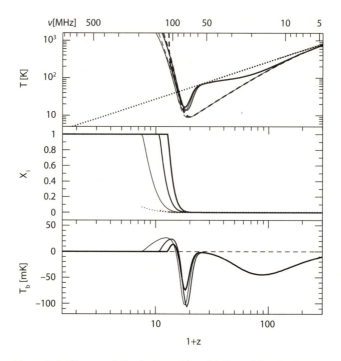

Figure 7.4. *Top panel:* Evolution with redshift z of the CMB temperature T_{CMB} (dotted curve), gas kinetic temperature T_k (dashed curve), and spin temperature T_S (solid curve). Following cosmological recombination at a redshift $z \sim 10^3$, the gas temperature tracks the CMB temperature $[\propto (1 + z)]$ down to a redshift $z \sim 200$ and then declines below it $[\propto (1 + z)^2]$, until the first X-ray sources (accreting black holes or exploding supernovae) heat it up well above the CMB temperature. The spin temperature of the 21-cm transition interpolates between the gas and CMB temperatures; initially it tracks the gas temperature through atomic collisions, then it tracks the CMB through radiative coupling, and eventually it tracks the gas temperature once again after the production of a cosmic background of UV photons that redshift into the Lyman-α

proportional to the neutral fraction. Also, if $T_S \gg T_{CMB}$, then the IGM is observed in emission at a level that is independent of T_S. On the other hand, if $T_S \ll T_{CMB}$, then the IGM is observed in absorption[88] at a level enhanced by a factor of T_{CMB}/T_S. As a result, a number of cosmic events are expected to leave observable signatures in the redshifted 21-cm line, as discussed below.

Figure 7.4 illustrates the mean IGM evolution for three examples in which reionization is completed at different redshifts, $z = 6.47$ (thin lines), $z = 9.76$ (medium thickness lines), and $z = 11.76$ (thick lines). The top panel shows the global evolution of the CMB temperature T_{CMB} (dotted curve), the gas kinetic temperature T_k (dashed curve), and the spin temperature T_S (solid curve). The middle panel shows the evolution of the ionized gas fraction and the bottom panel displays the mean 21 cm brightness temperature, T_b.

The prospect of studying reionization by mapping the distribution of atomic hydrogen across the Universe through its 21-cm spectral line has motivated several

---

resonance. *Middle panel:* Evolution of the gas fraction in ionized regions x_i (solid curve) and the ionized fraction outside these regions (due to diffuse X-rays) x_e (dotted curve). *Bottom panel:* Evolution of mean 21-cm brightness temperature T_b. The horizontal axis at the top provides the observed photon frequency for the different redshifts shown at the bottom. Each panel shows curves for three models in which reionization is completed at different redshifts: $z = 6.47$ (thin lines), $z = 9.76$ (medium thickness lines), and $z = 11.76$ (thick lines). Figure credit: J. Pritchard and A. Loeb, *Phys. Rev. D* **78**, 103511 (2008).

teams to design and construct arrays of low-frequency radio antennae. These teams plan to assemble arrays of thousands of dipole antennae and correlate their electric field measurements. Although the radio technology for the frequency range of interest is the same as used in past decades for TV or radio communication, the experiments have never been done before because computers were not sufficiently powerful to analyze and correlate the large volume of data produced by these arrays. The planned experiments include the Low Frequency Array,[89] the Murchison Wide-Field Array shown in figure 7.5,[90] the Primeval Structure Telescope,[91] the Precision Array to Probe the Epoch of Reionization,[92] and ultimately the Square Kilometer Array.[93] These low-frequency radio observatories will search over the next decade for redshifted 21-cm emission or absorption from redshifts $z \sim 6.5$–15, corresponding to observed wavelengths of 1.5–3.4 meters (comparable to the height of a person). Current observational projects in 21-cm cosmology are at the same status as was CMB research prior to the first statistical detection of the sky temperature fluctuations by the Cosmic Background Explorer (COBE) satellite.

Because the smallest angular size that can be resolved by a telescope is of order the observed wavelength divided by the telescope diameter, radio astronomy at wavelengths as large as a meter has remained relatively undeveloped. Production of resolved images even of large sources such as cosmological ionized bubbles requires telescopes which have a kilometer scale. It is much more cost effective to use a large array of thousands of simple antennas distributed

Figure 7.5. *Top:* Artificial illustration of the expected MWA experiment with 512 tiles (white spots) of 16 dipole antennae each, spread across an area of 1.5 km in diameter in the desert of western Australia. With a collecting area of 8,000 square meters, the array will be sensitive to 21-cm emission from cosmic hydrogen in the redshift range of $z = 6$–15 by operating in the radio frequency range of 80–300 MHz. *Bottom:* An actual image of one of the tiles. Image credits: Judd Bowman and Colin Lonsdale (2009).

over several kilometers, and to use computers to cross-correlate the measurements of the individual antennae and combine them effectively into a single large telescope. The new experiments are located mostly in remote sites, because the frequency band of interest overlaps with more mundane terrestrial telecommunications.*

Detection of the redshifted 21-cm signal is challenging. Relativistic electrons within our Milky Way galaxy produce synchrotron radio emission as they gyrate around the galactic magnetic field.[†] This results in a radio foreground that is larger than the expected reionization signal by at least a factor of ten thousand. But not all is lost. By shifting slightly in observed wavelength one is slicing the hydrogen distribution at different redshifts and hence one is seeing a different map of its bubble structure, but the synchrotron foreground remains nearly the same. Theoretical calculations demonstrate that it is possible to extract the signal from the epoch of reionization by subtracting the radio images of the sky at slightly different wavelengths. In approaching redshifted 21-cm observations, although the first inkling might be to consider the mean emission

---

*These experiments will bring a new capability to search for leakage of TV or radio signals from a distant civilization [A. Loeb and M. Zaldarriaga, *J. Cosmol. Astro-Part. Phys.* **1**, 20 (2007)]. Post World-War II leakage of radio signals from our civilization could have been detected by the same experiments out to a distance of tens of light years. Since our civilization produced its brightest radio signals for military purposes during the cold war, it is plausible that the brightest civilizations out there are the most militant ones. Therefore, if we do detect a signal, we had better not respond.

[†]For a pedagogical description of synchrotron radiation, see G. B. Rybicki and A. P. Lightman, *Radiative Processes in Astrophysics* (Wiley, New York, 1979), chap. 6.

signal in the bottom panel of figure 7.4 (and this is indeed the goal of some single-antenna experiments[94]), the signal is orders of magnitude fainter than the synchrotron foreground (see figure 7.6). Thus, most observers have focused on the expected characteristic variations in T_b, both with position on the sky and especially with frequency, which signifies redshift for the cosmic signal. The synchrotron foreground is expected to have a smooth frequency spectrum, so it is possible to isolate the cosmological signal by taking the difference in the sky brightness fluctuations at slightly different frequencies (as long as the frequency separation corresponds to the characteristic size of ionized bubbles). Large-scale patterns in the 21-cm brightness from reionization are driven by spatial variations in the abundance of galaxies; the 21-cm fluctuations reach a root-mean-square amplitude of $\sim 5\,\mathrm{mK}$ in brightness temperature on a scale of 10 comoving Mpc (figure 7.7). While detailed maps will be difficult to extract due to the foreground emission, a statistical detection of these fluctuations (through the power spectrum) is expected to be well within the capabilities of the first-generation experiments now being built. Current work suggests that the key information on the topology and timing of reionization can be extracted statistically.[95]

While numerical simulations of reionization are now reaching the cosmological box sizes needed to predict the large-scale topology of the ionized bubbles, they often do so at the price of limited small-scale resolution. These simulations cannot yet follow in any detail the formation of individual stars within galaxies, or the feedback from stars on the surrounding gas, such as heating by radiation

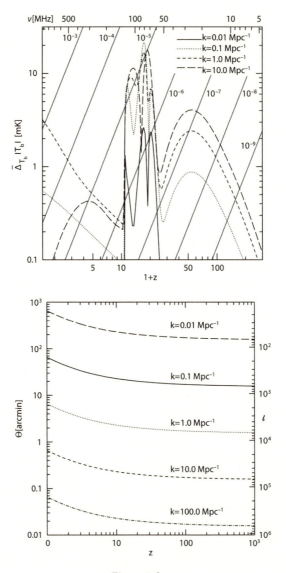

Figure 7.6.

or the piston effect of supernova explosions, which blow hot bubbles of gas enriched with the chemical products of stellar nucleosynthesis. The simulations cannot directly predict whether the stars that form during reionization are similar to the stars in the Milky Way and nearby galaxies or to the primordial $100 M_{\odot}$ stars. They also cannot determine whether feedback prevents low-mass dark matter halos from forming stars. Thus, models are needed to make it possible to vary all these astrophysical parameters of the ionizing sources and study the effect they have on the 21-cm observations.

The current theoretical expectations for reionization and for the 21-cm signal are based on rather large

---

Figure 7.6. *Facing page, top:* Predicted redshift evolution of the angle-averaged amplitude of the 21-cm power spectrum ($|\bar{\Delta}_{T_b}| = [k^3 P_{21\text{-cm}}(k)/2\pi^2]^{1/2}$) at comoving wavenumbers $k = 0.01$ (solid curve), 0.1 (dotted curve), 1.0 (short-dashed curve), 10.0 (long-dashed curve), and $100.0\,\text{Mpc}^{-1}$ (dot-dashed curve). In the model shown, reionization is completed at $z = 9.76$. The horizontal axis at the top shows the observed photon frequency at the different redshifts. The diagonal straight lines show various factors of suppression for the synchrotron galactic foreground, necessary to reveal the 21-cm signal. *Bottom:* Redshift evolution of the angular scale on the sky corresponding to different comoving wavenumbers, $\Theta = (2\pi/k)/d_A$. The labels on the right-hand side map angles to angular moments (often used to denote the multipole index of a spherical harmonics decomposition of the sky), using the approximate relation $\ell \approx \pi/\Theta$. Along the line of sight (the third dimension), an observed frequency bandwidth $\Delta\nu$ corresponds to a comoving distance of $\sim(1.8\,\text{Mpc})(\Delta\nu/0.1\,\text{MHz})[(1 + z)/10]^{1/2}$. Figure credit: J. Pritchard and A. Loeb, *Phys. Rev. D* **78**, 103511 (2008).

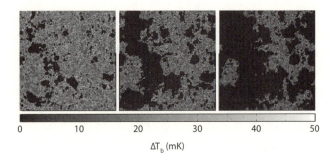

ΔT_b (mK)

Figure 7.7. Map of the fluctuations in the 21-cm brightness temperature on the sky, ΔT_b (mK), based on a numerical simulation which follows the dynamics of dark matter and gas in the IGM as well as the radiative transfer of ionizing photons from galaxies. The panels show the evolution of the signal in a slice of 140 comoving Mpc on a side, in three snapshots corresponding to the simulated volume being 25%, 50%, and 75% ionized. These snapshots correspond to the top three panels in figure 5.1. Since neutral regions correspond to strong emission (i.e., a high T_b), the 21-cm maps illustrate the global progress of reionization and the substantial large-scale spatial fluctuations in the reionization history. Figure credit: H. Trac, R. Cen, and A. Loeb, *Astrophys. J.* **689**, L81 (2009).

extrapolations from observed galaxies to deduce the properties of much smaller galaxies that formed at an earlier cosmic epoch. Considerable surprises are thus possible, such as an early population of quasars, or even unstable exotic particles that emitted ionizing radiation as they decayed. The forthcoming observational data in 21-cm cosmology should make the next decade a very exciting time.

It is of particular interest to separate signatures of the fundamental physics, such as the initial conditions from inflation and the nature of the dark matter and dark energy, from the astrophysics, involving phenomena related to star formation, which cannot be modeled accurately from first principles. This is particularly easy to do before the first galaxies formed ($z > 25$), at which time the 21-cm fluctuations are expected to simply trace the primordial power spectrum of matter density perturbations, which is shaped by the initial conditions from inflation and the dark matter. The same simplicity applies after reionization ($z < 6$)—when only dense pockets of self-shielded hydrogen (associated with individual galaxies) survive, and those behave as test particles and simply trace the matter distribution.[96] During the epoch of reionization, however, the 21-cm fluctuations are mainly shaped by the topology of ionized regions, and thus depend on uncertain astrophysical details involving star formation. However, even during this epoch, the imprint of deviations from the Hubble flow [i.e., peculiar velocity fluctuations **u** which are induced gravitationally by density fluctuations δ; see equations (3.3) and (3.4)], can in principle be used to separate the signatures of fundamental physics from the astrophysics.

Deviations from the smooth Hubble flow imprint a particular form of anisotropy in the 21-cm fluctuations caused by gas motions along the line of sight. This anisotropy, expected in any measurement of density based on a resonance line (or on any other redshift indicator), results from velocity compression. Consider a photon traveling along the line of sight that resonates with absorbing atoms at a particular point. In a uniform, expanding universe, the absorption optical depth encountered by this photon probes only a narrow strip of atoms, since the expansion of the universe makes all other neighboring atoms move with a relative velocity which takes them outside the narrow frequency width of the resonance line. If, however, there is a density peak near the resonating position, the increased gravity will reduce the expansion velocities around this point and bring more gas into the resonating velocity width. The associated Doppler effect is sensitive only to the line of sight component of the velocity gradient of the gas, and thus causes an observed anisotropy in the power spectrum even when all physical causes of the fluctuations are statistically isotropic. This anisotropy implies that the observed power spectrum in redshift space $P(\mathbf{k})$ depends on the angle between the line of sight and the $\mathbf{k}$-vector of a Fourier mode, not only on its amplitude k. This angular dependence allows to separate out the simple gravitational signature of density perturbations from the complex astrophysical effects of reionization, such as star and black hole formation, and feedback from supernovae and quasars.[97]

7.4 Observing Most of the Observable Volume

In general, cosmological surveys are able to measure the spatial power spectrum of primordial density fluctuations, $P(\mathbf{k})$, to a precision that is ultimately limited by Poisson statistics of the number of independent regions (the so-called cosmic variance). The fractional uncertainty in the amplitude of any Fourier mode of wavelength λ is given by $\sim 1/\sqrt{N}$, where N is the number of independent elements of size λ that fit within the survey volume. For the two-dimensional map of the CMB, N is the surveyed area of the sky divided by the solid angle occupied by a patch of area λ^2 at $z \sim 10^3$. 21-cm observations are advantageous because they access a three-dimensional volume instead of the two-dimensional surface probed by the CMB, and hence cover a larger number of independent regions in which the primordial initial conditions were realized. Moreover, the expected 21-cm power extends down to the pressure-dominated (Jeans) scale of the cosmic gas which is orders of magnitude smaller than the comoving scale at which the CMB anisotropies are damped by photon diffusion. Consequently, the 21-cm photons can trace the primordial inhomogeneities with a much finer resolution (i.e., many more independent pixels) than the CMB. Also, 21-cm studies promise to extend to much higher redshifts than existing galaxy surveys, thereby covering a much bigger fraction of the comoving volume of the observable Universe. At these high redshifts, small-scale modes are still in the perturbative (linear growth) regime where their statistical analysis is straightforward (chapter 3). Altogether, the above factors imply that the 21-cm mapping of cosmic hydrogen may potentially carry the largest number of bits of information about the initial conditions of our Universe compared to any other survey method in cosmology.[98]

The limitations of existing data sets (on which the cosmological parameters in table 3.1 are based) are apparent in figure 2.2, which illustrates the comoving volume of the Universe out to a redshift z as a function of z. State-of-the-art galaxy redshift surveys, such as the spectroscopic sample of luminous red galaxies (LRGs) in the Sloan Digital Sky Survey (SDSS), extend only out to $z \sim 0.3$ (only one-tenth of our horizon) and probe $\sim 0.1\%$ of the observable comoving volume of the Universe. Surveys of the 21-cm emission (or a large number of quasar skewers through the Lyman-α forest) promise to open a new window into the distribution of matter through the remaining 99.9% of the cosmic volume. These ambitious experiments might also probe the gravitational growth of structure through most of the observable Universe, and provide a new test of Einstein's theory of gravity across large scales of space and time.

8

EPILOGUE: FROM OUR GALAXY'S PAST TO ITS FUTURE

The previous chapters have been dedicated to the distant past of our cosmic ancestry, when the building blocks of the Milky Way galaxy were assembled in the form of the first galaxies. The laws of physics also allow us to forecast how our Galaxy and its environment will likely change in the future. Our cosmic perspective would be incomplete without a glimpse into this future.

8.1 End of Extragalactic Astronomy

Previous generations of scholars have occasionally wondered about the long-term future of the Universe or in Biblical Hebrew, the forecast for *acharit hayamim* ("the end of times"). For the first time in history, we now have a standard cosmological model that agrees with a large body of data about the past history of the Universe to an unprecedented precision. This model also makes scientific predictions about the future. Below let us summarize those predictions for the simplest version of the standard cosmological model with a steady cosmological constant.

Every time an American president delivers the "State of the Union" address, I imagine what it would be like to hear a supplementary comment about the "State of the Universe" surrounding the "Union." It is, of course, completely natural for a president to focus on issues affecting reelection on a four-year time scale while ignoring cosmological events that take billions of years to develop. Nevertheless, I find it amusing to imagine a brief statement about the bigger picture, especially when we have gained a significantly better understanding of our future in the cosmos than previous generations of astronomers had. And since cosmologists reached a major milestone of this quality over the past decade, let us consider below what this new understanding of the Universe might entail for our future.

As explained in chapters 2 and 6, cosmologists are able to observe what the Universe looked like in the past due to the finite speed of light. The light we observe today from a distant source must have been emitted at an earlier time in order for it to reach us today. Hence, by looking deeper into the Universe, we can see how it looked at earlier times. And if we are able to measure distances, we can also infer how the expansion rate of the Universe evolved with cosmic time. Naively, one would expect the expansion to have slowed down with time because of the gravitational attraction of cosmic matter. *But can this expectation be tested by observations through telescopes?*

A simple way to infer the distance of a street light with a 100-watt bulb is to measure how bright it appears at night. The 100 watts it produces are spread over a spherical surface area that increases in proportion to the square of the distance. Therefore, the observed flux of radiation (namely, the radiation energy per unit time per unit area) will get diluted as the distance squared and this can be used to calculate the distance [equations (6.1)–(6.4)]. Fortunately, exploding stars of a particular type (the so-called *type Ia supernovae*) provide a standardized "light bulb" that can be seen out to the edge of the Universe. More than a decade ago, two independent groups of observers inferred distances to a large number of such supernovae. Contrary to naive expectations, each discovered that the rate of cosmic expansion has been speeding up (i.e., *accelerating*) over the past five billion years.[99] As explained in chapter 2, Einstein's theory of gravity allows for an accelerating Universe if the mass density of the vacuum is greater than half that of matter* (which naturally occurs at late times as matter gets diluted by the cosmic expansion but the vacuum density stays constant). An unchanging vacuum, which would be equivalent to the so-called *cosmological constant*, naturally produces "repulsive" gravity that pushes galaxies away from each other at an ever increasing rate. With this discovery in mind, what should the president say about the "state of our Universe"?

---

*It is possible, in principle, to measure the cosmic acceleration in real time in terms of the change in the redshifts of the Lyman-α forest; see A. Loeb, *Astrophys. J.* **499**, L111 (1998). Other methods appear impractical; see, e.g. A. Sandage, *Astrophys. J.* **136**, 319 (1962). There are plans to measure this subtle shift in the Lyman-α forest with future large telescopes; see J. Liske et al., *Mon. Not. R. Astron. Soc.* **386**, 1192 (2008).

To put it bluntly: *the state of our Universe is not looking good for future observers.* If the Universe is indeed dominated by a cosmological constant and if the corresponding vacuum density will remain constant as the Universe ages by another factor of ten, then all galaxies outside our immediate vicinity will be pushed out of our horizon by the accelerated expansion of space (see figure 8.1). In a short paper published in 2002, entitled "The Long-Term Future of Extragalactic Astronomy," I showed that in fact all galaxies beyond a redshift of $z = 1.8$ are already outside our horizon right now (see figure 8.1). The light that we receive from them at this time was emitted a long while ago and does not represent their current state. If this light encodes some message from a distant civilization, then any reply we send back will be lost in space and never reach its target. Within a finite time, the accelerated expansion of space moves any distant galaxy away from us at a speed that exceeds the speed of light. Even light (which travels at the fastest speed by which material particles can propagate) is ultimately unable to bridge across the inflating gap that is opened between distant galaxies and us by the accelerated cosmic expansion. The situation is similar to observing a friend cross the event horizon of a black hole.[100] From a distance, we would see the image of our adventurous friend heading toward a black hole, getting fainter, and eventually freezing at some final snapshot, as that friend appears to be hovering just above the horizon. The frozen image will represents the snapshot of the friend as he/she crosses the horizon in his/her own rest frame. Beyond that point in time, no information can be retrieved about the whereabouts of the friend in classical general relativity.

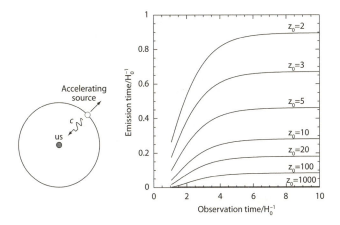

Figure 8.1. The accelerated cosmic expansion of space pulls distant galaxies apart from us at a speed that increases with time. When the speed of a galaxy relative to us exceeds the speed of light, the image of that galaxy freezes and fades away. No additional information can be gathered on the subsequent history of that galaxy, and no later communication is possible with its inhabitants. The figure on the right shows the time at which light is emitted by a distant galaxy as a function of the future time at which it will be observed. Time is measured in units of $t_H = H_0^{-1} = 14\,\text{Gyr}$, and the current cosmic time is $t_0 = 0.96 H_0^{-1}$, where H_0 is the present-day Hubble constant. For any currently measured redshift z_0 of a source, there is a maximum intrinsic age up to which we can see that source evolving even if we continue to monitor it indefinitely. Figure credit: A. Loeb, *Phys. Rev. D* **65**, 047301 (2002).

Since the image of each galaxy will eventually freeze at some finite time in its own rest frame, there is a limited amount of information we can collect about each distant galaxy. In particular, we will never be able to study the evolution of a galaxy beyond some finite age in its own frame of reference. The farther away a galaxy is, the earlier its image freezes, and the less information we get about its late evolution.

Of course, bound systems held together by a force stronger than the repulsive force of the cosmological constant do not participate in the cosmic expansion. This includes electrons bound to atoms by the electromagnetic force, planets bound to the Sun, as well as galaxies bound together by a stronger gravitational pull than the cosmic repulsion.

8.2 Milky Way + Andromeda = Milkomeda

Which galaxies will remain gravitationally bound to the Milky Way and which will fly away? Figure 8.2 shows the results from a computer simulation of the future evolution of all galaxies within 130 million light years of the Milky Way. Using a realization of the local distribution of galaxies based on existing observational data, the simulation predicts that all galaxies outside the local group of galaxies within a distance of 3 million light years are not bound to us and will be pushed out of our view by the cosmic expansion. The gas between galaxies will get extremely cold, but remain mostly ionized, as it is today.[101]

How will the local group of galaxies look in the distant future? The two major constituents of the local group of galaxies, namely, our own Milky Way galaxy and its nearest neighbor, the Andromeda galaxy (figure 8.3), are on a collision course. The Doppler shift of spectral lines from Andromeda indicates that Andromeda and the Milky

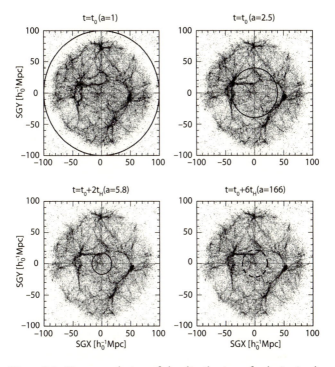

Figure 8.2. Future evolution of the distribution of galaxies in the vicinity of the Milky Way galaxy. The cosmic expansion was taken out from this visualization by using comoving coordinates that expand together with the Universe. Shown are galaxies projected in a slab of thickness 43 Mpc (the so-called supergalactic XY plane) at times $t = t_0$, $t_0 + t_H$, $t_0 + 2t_H$, and $t_0 + 6t_H$ where t_0 is the present time and $t_H = 14$ Gyr (corresponding to a cosmic scale factor $a = 1.0$, 2.5, 5.8, and 166), from top left to bottom right. The thick solid circle in each panel indicates the physical radius of 143 Mpc. In the bottom right panel, the Universe has expanded so much that this circle is no longer visible. Instead, we show the event horizon at a physical radius of 5.1 Gpc as the thick dashed circle. Figure credit: K. Nagamine and A. Loeb, *New Astronomy* **8**, 439 (2003).

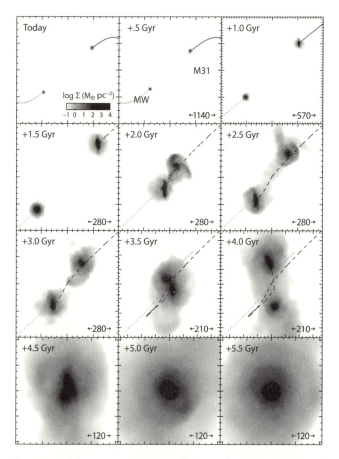

Figure 8.3. The merger product of the Milky Way (MW) and its nearest neighbor, the Andromeda (M31) galaxy, will be the only object visible to us in the night sky once the Universe ages by another factor of ten. The set of images show the forecast for the evolution of this future collision based on a computer simulation. The time sequence of the projected density of stars is shown. Andromeda is the larger of the two galaxies and begins the simulation in the upper right. The Milky Way begins on the edge

Way are moving toward each other at a radial velocity $\sim 100 \, \mathrm{km \, s^{-1}}$, but provides no information on the transverse speed. Other clues suggests that the transverse speed is modest;[102] the two galaxies appear to be gravitationally bound and will inevitably lose their transverse speed through dynamical friction on the dark matter around them.

For the most likely parameters of the local group, the merger of the Milky Way and Andromeda will take place within a few billion years (see figure 8.3). Since the lifetime of the Sun is somewhat longer,[103] this event could potentially be observable for future inhabitants of the solar system. *What will happen to the solar system during the merger?* Figure 8.4 shows that the Sun will most likely be kicked out to a distance that is a few times larger than its current separation from the center of the Milky Way. During the interaction, the Milky Way might appear as an external galaxy to observers within the solar system. Also, the night sky will change. Currently, the thin disk of the Milky Way in which we reside appears as a strip of stars on the sky. These stars will be scattered by the merger into a spheroid-like distribution. The merger product of the collision between the Milky Way and Andromeda will make

---

of the image in the lower left. Panels have varying spatial scales as specified by the label, in kpc, on the lower right of each panel. The simulation time, in Gyr relative to today, appears on the top left label of each panel. The trajectories of the Milky Way and Andromeda are depicted by the solid lines. Image credit: T. J. Cox and A. Loeb, *Mon. Not. R. Astron. Soc.* **386**, 461 (2008). This is the only paper in my publication record that has a chance of being cited in a few billion years.

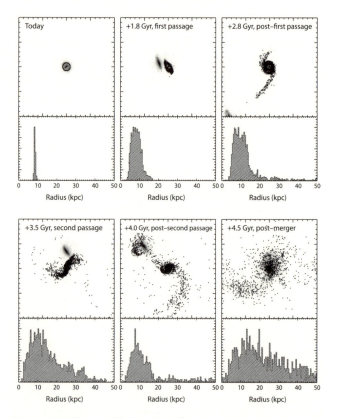

Figure 8.4. The possible location of the Sun during various stages of the merger between the Milky Way and Andromeda. The top component of each panel shows all stellar particles in the simulation that have a present-day galacto-centric radius of 8 ± 0.1 kpc as dots, and tracks their position into the future. The bottom component displays a histogram for the probability of having different radial distances of these particles from the center of the Milky Way. It is difficult to forecast which of these particles is the Sun, because the orbital time of the Sun around the galaxy is short compared to the merger time. Figure credit: T. J. Cox and A. Loeb, *Mon. Not. R. Astron. Soc.* **386**, 461 (2008).

a new galaxy, which I labeled in our paper on this topic as "Milkomeda." The spheroidal shape of Milkomeda is not unusual, as it characterizes a major class of galaxies called "elliptical galaxies." Presumably, many of these galaxies in the present-day Universe formed out of mergers between galactic disks at earlier cosmic times. Ultimately, only Milkomeda will remain visible to future solar system astronomers.

The current state of an accelerating Universe makes the study of cosmology a transient episode in our long-term scientific endeavor. We had better observe the Universe in the next tens of billions of years and document our findings for the benefit of future scientists, who will not be able to do so. It would also seem prudent for funding agencies to allocate their funds for cosmological observations before time runs out. Once cosmological observations become pointless, the funds could be taken away from this line of research and be dedicated instead to solving our local political problems. At that time, the future president, in giving his or her State of the Union address, will be fully justified in not having anything to say about the empty state of our Universe at large.

APPENDIX: USEFUL NUMBERS

Fundamental Constants

Newton's constant (G)	$= 6.67 \times 10^{-8} \, \text{cm}^3 \, \text{g}^{-1} \, \text{s}^{-2}$
Speed of light (c)	$= 3.00 \times 10^{10} \, \text{cm s}^{-1}$
Planck's constant (h)	$= 6.63 \times 10^{-27} \, \text{erg s}$
Electron mass (m_e)	$= 9.11 \times 10^{-28} \, \text{g}$
	$\equiv 511 \, \text{keV}/c^2$
Electron charge (e)	$= 4.80 \times 10^{-10} \, \text{esu}$
Proton mass (m_p)	$= 1.67 \times 10^{-24} \, \text{g}$
	$= 938.3 \, \text{MeV}/c^2$
Boltzmann's constant (k_B)	$= 1.38 \times 10^{-16} \, \text{erg K}^{-1}$
Stefan-Boltzmann constant (σ)	$= 5.67 \times 10^{-5} \, \text{erg cm}^{-2} \, \text{s}^{-1} \text{K}^{-4}$
Radiation constant (a)	$= 7.56 \times 10^{-15} \, \text{erg cm}^{-3} \, \text{K}^{-4}$
Thomson cross section (σ_T)	$= 6.65 \times 10^{-25} \, \text{cm}^2$

Astrophysical Numbers

Solar mass $(M_\odot)$	$= 1.99 \times 10^{33} \, \text{g}$
Solar radius $(R_\odot)$	$= 6.96 \times 10^{10} \, \text{cm}$

Solar luminosity ($L_\odot$)	$= 3.9 \times 10^{33}$ erg s^{-1}
Hubble constant today (H_0)	$= 100 h_0$ km s^{-1} Mpc^{-1}
Hubble time (H_0^{-1})	$= 3.09 \times 10^{17} h_0^{-1}$ s
	$= 9.77 \times 10^9 h_0^{-1}$ yr
	$\equiv 3 h_0^{-1}$ Gpc$/c$
critical density (ρ_c)	$= 1.88 \times 10^{-29} h_0^2$ g cm^{-3}
	$= 1.13 \times 10^{-5} h_0^2 m_p$ cm^{-3}

Unit Conversions

1 parsec (pc)	$= 3.086 \times 10^{18}$ cm
1 kiloparsec (kpc)	$= 10^3$ pc
1 megaparsec (Mpc)	$= 10^6$ pc
1 gigaparsec (Gpc)	$= 10^9$ pc
1 astronomical unit (AU)	$= 1.5 \times 10^{13}$ cm
1 year (yr)	$= 3.16 \times 10^7$ s
1 light year (ly)	$= 9.46 \times 10^{17}$ cm
1 eV	$= 1.60 \times 10^{-12}$ ergs
	$\equiv 11{,}604$ K $\times k_B$
1 erg	$= 10^7$ J
Photon wavelength ($\lambda = c/\nu$)	$= 1.24 \times 10^{-4}$ cm (photon energy$/1$ eV)$^{-1}$
1 nanojansky (nJy)	$= 10^{-32}$ erg cm^{-2} s^{-1} Hz^{-1}
1 angstrom (Å)	$= 10^{-8}$ cm
1 micron (μm)	$= 10^{-4}$ cm
1 km s^{-1}	$= 1.02$ pc per million years
1 arcsecond ($''$)	$= 4.85 \times 10^{-6}$ radians
1 arcminute ($'$)	$= 60''$
1 degree ($^\circ$)	$= 3.6 \times 10^{3\,''}$
1 radian	$= 57.3^\circ$

NOTES

Chapter 2. Standard Cosmological Model

1. P. de Bernardis et al., *Nature* **404**, 955, (2000); S. Hanany et al., *Astrophys. J.* **545**, L5 (2000); A. D. Miller et al., *Astrophys. J.* **524**, L1 (1999).

2. See, e.g., P.J.E. Peebles, *Principles of Physical Cosmology* (Princeton University Press, Princeton, NJ, 1993), in particular pp. 62–65.

3. For advanced reading, see V. Mukhanov, *Physical Foundations of Cosmology* (Cambridge University Press, Cambridge, 2005).

4. http://public.web.cern.ch/public/en/LHC/LHC-en.html

Chapter 3. The First Gas Clouds

5. A. Loeb, A. Ferrara, and R. S. Ellis, *First Light in the Universe*, Saas-Fee Advanced Course 36 (Springer, New York, 2008), and references therein.

6. Z. Haiman, A. A. Thoul, and A. Loeb, *Astrophys. J.* **464**, 523 (1996).

7. R. Barkana and A. Loeb, *Astrophys. J.* **523**, 54 (1999).

8. A. Loeb and M. Zaldarriaga, *Phys. Rev. D* **71**, 103520 (2005).

9. R. Barkana and A. Loeb, *Astrophys. J.* **531**, 613 (2000), and references therein.

10. W. H. Press and P. Schechter, *Astrophys. J.* **187**, 425 (1974).

11. J. R. Bond, S. Cole, G. Efstathiou, and N. Kaiser, *Astrophys. J.* **379**, 440 (1991).

12. C. G. Lacey and S. Cole, *Mon. Not. R. Astron. Soc.* **262**, 627 (1993).

13. Z. Haiman, M. J. Rees, and A. Loeb, *Astrophys. J.* **476**, 458 (1997).

Chapter 4. The First Stars and Black Holes

14. V. Bromm and R. B. Larson, *Annu. Rev. Astron. Astrophys.* **42**, 79 (2004), and references therein.

15. J.S.B Wyithe and A. Loeb, *Nature* **441**, 322 (2006).

16. J. A. Muñoz, P. Madau, A. Loeb, and J. Diemand, *Mon. Not. R. Astron. Soc.*, in press (2009), http://arxiv.org/abs/0905.4744, and references therein.

17. M. J. Turk, T. Abel, and B. O'Shea, *Science* **325**, 601 (2009); A. Stacy, T. H. Greif, and V. Bromm, *Mon. Not. R., Astron. Soc.*, submitted (2009), http://arxiv.org/abs/0908.0712, and references therein.

18. M. Dijkstra and A. Loeb, *Mon. Not. R. Astron. Soc.* **391**, 457 (2008).

19. V. Bromm and A. Loeb, *New Astron.* **9**, 353 (2004).

20. R. E. Pudritz, *Science* **295**, 68 (2002), and references therein.

21. E. Salpeter, *Astrophys. J.* **121**, 161 (1955).

22. M. J. Rees, *Mon. Not. R. Astron. Soc.* **176**, 483 (1976).

23. V. Bromm and A. Loeb, *Nature* **425**, 812 (2003).

24. S. R. Furlanetto and A. Loeb, *Astrophys. J.* **556**, 619 (2001).

25. S. R. Furlanetto and A. Loeb, *Astrophys. J.* **588**, 18 (2003).

26. J. Miralda-Escudé and M. Rees, *Astrophys. J.* **478**, L57 (2007).

27. K. Wood and A. Loeb, *Astrophys. J.* **545**, 86 (2000).

28. M. Fukugita, C. J. Hogan, and P.J.E. Peebles, *Astrophys. J.* **503**, 518 (1998).

29. See, e.g., R. Barkana and A. Loeb, *Astrophys. J.* **539**, 20 (2000); D. P. Stark, A. Loeb, and R. S. Ellis, *Astrophys. J.* **668**, 627 (2007), and references therein.

30. H. J. Mo and S.D.M. White, *Mon. Not. R. Astron. Soc.* **336**, 112 (2002).

31. A. Frebel, J. L. Johnson, and V. Bromm, *Astrophys. J.* **392**, L50 (2009), and references therein.

32. M. C. Begelman, R. D. Blandford, and M. J. Rees, *Rev. Mod. Phys.* **56**, 255 (1984).

33. See S. L. Shapiro and S. A. Teukolsky, *Black Holes, White Dwarfs, and Neutron Stars* (Wiley, New York, 1982), chap. 14.

34. J.S.B. Wyithe and A. Loeb, *Astrophys. J.* **595**, 614 (2003).

35. See A. Loeb, http://arxiv.org/abs/0909.0261 (2009), and references therein.

36. F. Pretorius, *Phys. Rev. Lett.* **95**, 121101 (2005); M. Campanelli et al., *Phys. Rev. Lett.* **96**, 111101 (2006); J. Baker et al., *Phys. Rev. Lett.* **96**, 111102 (2006).

37. A. Loeb, *Phys. Rev. Lett.* **99**, 041103 (2007).

38. L. Blecha and A. Loeb, *Mon. Not. R. Astron. Soc.* **390**, 1311 (2008); T. Tanaka and Z. Haiman, *Astrophys. J.* **696**, 1798 (2009).

39. K. Gebhardt et al., *Astrophys. J.* **539**, L13 (2000); L. Ferrarese and D. Merritt, *Astrophys. J.* **539**, L9 (2000). These two competing papers appeared simultaneously since I suggested the topic to their lead authors at the same time.

40. J. Magorrian et al., *Astron. J.* **115**, 2285 (1998).

41. J. Silk and M. J. Rees, *Astron. Astrophys.* **331**, L1 (1998).

42. M. Milosavljević and A. Loeb, *Astrophys. J.* **604**, L45 (2004).

43. M. Dietrich and F. Hamann, *Rev. Mex. Astron. Astrof. Conf. Ser.* **32**, 65 (2008).

44. See, e.g., J.S.B. Wyithe and A. Loeb, *Astrophys. J.* **595**, 614 (2003); P. F. Hopkins and L. Hernquist, *Astrophys. J.* **698**, 1550 (2009).

45. R. Genzel and V. Karas, *IAU Symp.* **238**, 173 (2007); A. Ghez et al., *Astrophys. J.* **689**, 1044 (2008).

46. For recent reviews about GRBs, see P. Mészáros, *AIP Conf. Proc.* **924**, 3 (2007) and *Rep. Prog. Phys.* **69**, 2259 (2006); T. Piran, *Nuovo Cimento B* **121**, 1039 (2006) and *Rev. Mod. Phys.* **76**, 1143 (2005); and N. Gehrels, E. Ramirez-Ruiz, and D. B. Fox, *Annu. Rev. Astron. Astrophys.* **47**, 567 (2009).

47. W. Zhang, S. E. Woosley, and A. I. MacFadyen, *J. Phys. Conf. Ser.* **46**, 403 (2006).

48. http://swift.gsfc.nasa.gov/

49. N. R. Tanvir et al., *Nature* **461**, 1254 (2009); R. Salvaterra et al., *Nature* **461**, 1258 (2009).

Chapter 5. The Reionization of Cosmic Hydrogen by the First Galaxies

50. J. Dunkley et al., *Astrophys. J. Suppl.* **180**, 306 (2009).

51. C.-A. Faucher-Giguère, A. Lidz, L. Hernquist, and M. Zaldarriaga, *Astrophys. J.* **682**, L9 (2008); J.S.B. Wyithe and A. Loeb, *Astrophys. J.* **586**, 693 (2003).

52. B. Strömgren, *Astrophys. J.* **89**, 526 (1939).

53. P. R. Shapiro and M. L. Giroux, *Astrophys. J.* **321** L107 (1987).

54. R. Barkana and A. Loeb, *Phys. Rep.* **349**, 129 (2001), and references therein.

55. J.S.B. Wyithe and A. Loeb, *Nature* **427**, 815 (2004); *Astrophys. J.* **610**, 117 (2004).

56. N. Gnedin, *Astrophys. J.* **535**, 530 (2000).

57. R. Barkana and A. Loeb, *Astrophys. J.* **609**, 474 (2004).

58. S. R. Furlanetto, M. Zaldarriaga, and L. Hernquist, *Astrophys. J.* **613**, 1 (2004).

59. O. Zahn et al., *Astrophys. J.* **654**, 12 (2007).

60. N. Kaiser, *Astrophys. J.* **284**, L9 (1984).

61. J.S.B. Wyithe and A. Loeb, *Nature* **432**, 194 (2004).

Chapter 6. Observing the First Galaxies

62. See, e.g., overview in R. Barkana and A. Loeb, *Phys. Rep.* **349**, 125 (2000), chap. 8; and also Z. Haiman and A. Loeb, *Astrophys. J.* **483**, 21 (1997).

63. For an overview of the current observational status, see R. S. Ellis, http://arxiv.org/abs/astro-ph/0701024 (2007).

64. R. I. Thompson, *Astrophys. J.* **596**, 748 (2003).

65. R. J. Bouwens, G. D. Illingworth, M. Franx, and H. Ford, *Astrophys. J.* **686**, 230 (2008).

66. See, e.g., N. Kashikawa et al., *Astrophys. J.* **648**, 7 (2006); S. Dawson et al., *Astrophys. J.* **671**, 1227 (2007).

67. See, e.g., H. Yan et al., *Astrophys. J.* **651**, 24 (2006); J. E. Rhoads et al., *Astrophys. J.* **697**, 942 (2009); M. Ouchi et al., *Astrophys. J.* **706**, 1136 (2009).

68. K. Ota et al., *Astrophys. J.* **677**, 12 (2008).

69. L. Bradley et al., *Astrophys. J.* **678**, 647 (2008).

70. D. Stark et al., *Astrophys. J.* **663**, 10 (2007); R. J. Bouwens et al., *Astrophys. J.* **690**, 1764 (2009).

71. B. Ciardi and A. Loeb, *Astrophys. J.* **540**, 687 (2000).

72. R. Barkana and A. Loeb, *Astrophys. J.* **601**, 64 (2004).

73. V. Bromm and A. Loeb, *Astrophys. J.* **642**, 382 (2006).

74. See, e.g., R. Barkana and A. Loeb, *Astrophys. J.* **531**, 613 (2000); **539**, 20 (2000).

75. V. Bromm, R. P. Kudritzki, and A. Loeb, *Astrophys. J.* **552**, 464 (2001).

76. http://www.eso.org/sci/facilities/eelt/

77. http://www.gmto.org/
78. http://www.tmt.org/
79. J.S.B. Wyithe and A. Loeb, *Mon. Not. R. Astron. Soc.* **375**, 1034 (2007).
80. http://almaobservatory.org/

Chapter 7. Imaging the Diffuse Fog of Cosmic Hydrogen

81. P.A.G. Scheuer, *Nature* **207**, 963 (1965).
82. J. E. Gunn and B. A. Peterson, *Astrophys. J.* **142**, 1633 (1965).
83. J. Miralda-Escudé, *Astrophys. J.* **501**, 15 (1998).
84. For further reading on 21-cm cosmology, see S. R. Furlanetto, S. P. Oh, and F. H. Briggs, *Phys. Rep.* **433**, 181 (2006), and references therein.
85. A. Loeb and M. Zaldarriaga, *Phys. Rev. Lett.* **92**, 211301 (2004).
86. S. A. Wouthuysen, *Astron. J.* **57**, 31 (1952); G. B. Field, *Proc. IRE* **46**, 240 (1958).
87. P. Madau, A. Meiksin, and M. J. Rees, *Astrophys. J.* **475**, 429 (1997).
88. D. Scott and M. J. Rees, *Mon. Not. R. Astron. Soc.* **247**, 510 (1990).
89. http://www.lofar.org/
90. http://www.haystack.mit.edu/ast/arrays/mwa/site/index.html
91. http://arxiv.org/abs/astro-ph/0502029
92. http://arxiv.org/abs/0904.2334
93. http://www.skatelescope.org
94. See, e.g., J. D. Bowman, A.E.E. Rogers, and J. N. Hewitt, *Astrophys. J.* **676**, 1 (2008).
95. See, e.g., A. Lidz, O. Zahn, M. McQuinn, M. Zaldarriaga, and L. Hernquist, *Astrophys. J.* **680**, 962 (2008), and references therein.

96. J.S.B. Wyithe and A. Loeb, *Mon. Not. R. Astron. Soc.* **383**, 1195 (2008).

97. R. Barkana and A. Loeb, *Astrophys. J.* **624**, L65 (2005).

98. A. Loeb and J.S.B. Wyithe, *Phys. Rev. Lett.* **100**, 161301 (2008); Y. Mao et al., *Phys. Rev. D* **78**, 023529 (2008).

Chapter 8. Epilogue: From Our Galaxy's Past to Its Future

99. A. Riess et al., *Astron. J.* **116**, 1009 (1998); S. Perlmutter et al., *Astrophys. J.* **517**, 565 (1999).

100. For further reading on black hole horizons, see K. S. Thorne, *Black Holes and Time Warps* (W. W. Norton & Company, New York, 1994).

101. K. Nagamine and A. Loeb, *New Astronomy* **9**, 573 (2004).

102. R. P. van der Marel and P. Guhathakurta, *Astrophys. J.* **678**, 187 (2008); A. Loeb, M. Reid, A. Brunthaler, and H. Falcke, *Astrophys. J.* **633**, 894 (2005).

103. P. Schröder and R. Connon Smith, *Mon. Not. R. Astron. Soc.* **386**, 155 (2008).

RECOMMENDED FURTHER READING

Cosmology

Kolb, E. W., and M. S. Turner, *The Early Universe*, Addison-Wesley, Reading, MA (1990).

Mukhanov, V., *Physical Foundations of Cosmology*, Cambridge University Press, Cambridge (2005).

Padmanabhan, T., *Structure Formation in the Universe*, Cambridge University Press, Cambridge (1993).

Peebles, P.J.E., *Principles of Physical Cosmology*, Princeton University Press, Princeton, NJ (1993).

Introduction to Astrophysics

Maoz, D., *Astrophysics in a Nutshell*, Princeton University Press, Princeton, NJ (2007).

Schneider, P., *Extragalactic Astronomy and Cosmology*, Springer-Verlag, Berlin (2006).

Radiative and Collisional Processes

Osterbrock, D. E., and G. J. Ferland, *Astrophysics of Gaseous Nebulae and Active Galactic Nuclei* (2nd ed.), University Science Books, Herndon, VA (2006).

Rybicki, G. B., and A. P. Lightman, *Radiative Processes in Astrophysics*, Wiley-Interscience, New York (1979).

Compact Objects

Peterson, B. M., *An Introduction to Active Galactic Nuclei*, Cambridge University Press, Cambridge (1997).

Shapiro, S. L., and S. A. Teukolsky, *Black Holes, White Dwarfs, and Neutron Stars: The Physics of Compact Objects*, Wiley-Interscience, New York (1983).

Galaxies

Binney, J., and M. Merrifield, *Galactic Astronomy*, Princeton University Press, Princeton, NJ (1998).

Binney, J., and S. Tremaine, *Galactic Dynamics* (2nd ed.), Princeton University Press, Princeton, NJ (2008).

GLOSSARY

Baryons: strongly interacting particles made of three quarks, such as the proton and the neutron from which atomic nuclei are made. Baryons carry most of the mass of ordinary matter, since the proton and neutron masses are nearly two thousand times higher than the electron mass. Electrons and neutrinos are called **leptons** and are subject to only the electromagnetic, gravitational, and weak interactions.

Big Bang: the moment in time when the expansion of the Universe started. We cannot reliably extrapolate our history before the Big Bang because the densities of matter and radiation diverge at that time. A transition through the Big Bang could be described only by a future theory that will unify quantum mechanics and gravity.

Blackbody radiation: the radiation obtained in complete thermal equilibrium with matter of some fixed temperature. The intensity of the radiation as a function of photon wavelength is prescribed by the Planck spectrum. The best experimental confirmation of this spectrum was obtained by the COBE satellite measurement of the cosmic microwave background (CMB).

Black hole: a region surrounded by an **event horizon** from which no particle (including light) can escape. A black hole is the end product from the complete gravitational collapse of a material object, such as a massive star or a gas cloud. It is characterized only by its mass, charge, and spin (similarly to elementary particles).

Cosmic inflation: an early phase transition during which the cosmic expansion accelerated, and the large-scale conditions of the present-day Universe were produced. These conditions include the large-scale homogeneity and isotropy, the flat global geometry, and the spectrum of the initial density fluctuations, which were all measured with exquisite precision over the past two decades.

Cosmic microwave background (CMB): the relic thermal radiation left over from the opaque hot state of the Universe before cosmological recombination.

Cosmological constant (dark energy): the mass (energy) density of the vacuum (after all forms of matter or radiation are removed). This constituent introduces a repulsive gravitational force that accelerates the cosmic expansion. The cosmic mass budget is observed to be dominated by this component at the present time (as it carries more than twice the combined mass density of ordinary matter and dark matter).

Cosmological principle: a combination of two constraints which describe the Universe on large scales: (i) homogeneity (same conditions everywhere), and (ii) isotropy (same conditions in all directions).

Cosmology: the scientific study of the properties and history of the Universe. This research area includes **observational** and **theoretical** subfields.

Dark matter: a mysterious dark component of matter which reveals its existence only through its gravitational influence and leaves no other clue about its nature. The nature of the dark matter is unknown, but searches are under way for an associated weakly interacting particle.

Galaxy: an object consisting of a luminous core made of stars or cold gas surrounded by an extended halo of dark matter. The stars in galaxies are often organized in either a disk (often with spiral arms) or ellipsoidal configurations, giving rise to **disk** (spiral) or **elliptical** (spheroidal) galaxies, respectively. Our own Milky Way galaxy is a disk galaxy with a central spheroid. Since we observe our galaxy from within, its disk stars appear to cover a strip across the sky.

Gamma-ray burst (GRB): a brief flash of high-energy photons which is often followed by an afterglow of lower-energy photons on longer timescales. Long-duration GRBs (lasting more than a few seconds) are believed to originate from relativistic jets which are produced by a black hole after the gravitational collapse of the core of a massive star. They are often followed by a rare (type Ib/c) supernova associated with the explosion of the parent star. Short-duration GRBs are thought to originate also from the coalescence of compact binaries which include two neutron stars or a neutron star and a black hole.

Hubble parameter $H(t)$: the ratio between the cosmic expansion speed and distance within a small region in a homogeneous and isotropic Universe. Formulated empirically by Edwin Hubble in 1929 based on local observations of galaxies. H is time-dependent but spatially constant at any given time. The inverse of the Hubble parameter, also called the **Hubble time**, is of order the age of the Universe.

Hydrogen: a proton and an electron bound together by their mutual electric force. Hydrogen is the most abundant element in the Universe (accounting for $\sim 76\%$ of the primordial mass budget of ordinary matter), followed by helium ($\sim 24\%$), and small amounts of other elements.

Jeans mass: the minimum mass of a gas cloud required in order for its attractive gravitational force to overcome the repulsive pressure force of the gas. First formulated by the physicist James Jeans.

Linear perturbation theory: a theory describing the gravitational growth of small-amplitude perturbations in the cosmic matter density, by expanding the fundamental dynamical equations to leading order in the perturbation amplitude.

Lyman-α transition: a transition between the ground state ($n = 1$) and the first excited level ($n = 2$) of the hydrogen atom. The associated photon wavelength is $1216\,\text{Å}$.

Neutron star: a star made almost exclusively of neutrons, formed as a result of the gravitational collapse of the core of a massive star progenitor. A neutron star has a mass comparable to that of the Sun and a mass density comparable to that of an atomic nucleus.

Quasar: a bright compact source of radiation which is powered by the accretion of gas onto a massive black hole. The relic (dormant) black holes from quasar activity at early cosmic times are found at the centers of present-day galaxies.

Recombination of hydrogen: the assembly of hydrogen atoms out of free electrons and protons. Cosmologically, this process occurred 0.4 million years after the Big Bang at a redshift of $\sim 1.1 \times 10^3$, when the temperature first dipped below $\sim 3 \times 10^3$ K.

Reionization of hydrogen: the breakup of hydrogen atoms, left over from cosmological recombination, into their constituent electrons and protons. This process took place hundreds of millions of years after the Big Bang, and is believed to have resulted from the UV emission by stars in the earliest generation of galaxies.

Star: a dense, hot ball of gas held together by gravity and powered by nuclear fusion reactions. The closest example is the Sun.

Supernova: the explosion of a massive star after its core has consumed its nuclear fuel.

21-cm transition: a transition between the two states (up or down) of the electron spin relative to the proton spin in a hydrogen atom. The associated photon wavelength is 21 cm.

INDEX